Mobilities, Mobility Justice and Social Justice

This collection investigates the relationship between mobilities and social justice to develop the concept of *mobility justice*.

Two introductory chapters outline how social justice concepts can strengthen analyses of mobility as socially structured movement in particular fields of power, what new justice-related questions arise by considering uneven mobilities through a social justice frame, and what a 'mobile ontology' contributes to understandings of justice in relation to 21st-century social relations. In 15 subsequent chapters, authors analyze the material infrastructures that configure mobilities and co-constitute injustice, the justice implications of 'more-than-human' movements of food and animals, and mobility-related injustices produced in relation to institutional acts of governance and through micro-scale embodied relations of race, gender, class and sexuality that shape the uneven freedom of human bodily movements.

The volume brings numerous scales, types and facets of mobility into conversation with multiple approaches to social justice in order to theorize mobility justice and reimagine social justice as a *mobile* concept appropriate for analyzing the effects and ethics of contemporary life. It is aimed at scholars and upper-level students in the interdisciplinary fields of critical mobilities and social justice, especially from disciplinary locations in geography, sociology, philosophy, transport planning, anthropology, and design and urban studies.

Nancy Cook is Associate Professor of Sociology at Brock University, Canada. Her research in northern Pakistan focuses on mobility justice, mobility disaster, and the gendered constitution of mobility.

David Butz is Professor of Geography at Brock University, Canada, and Editor-in-Chief of *Studies in Social Justice*. He studies the social implications of road construction and associated (im)mobilities in rural northern Pakistan.

Mobilities, Mobility Justice and Social Justice

Edited by
Nancy Cook and David Butz

Routledge
Taylor & Francis Group

LONDON AND NEW YORK

First published 2019
by Routledge
2 Park Square, Milton Park, Abingdon, Oxon OX14 4RN
52 Vanderbilt Avenue, New York, NY 10017, USA

First issued in paperback 2020

Routledge is an imprint of the Taylor & Francis Group, an informa business

British Library Cataloguing-in-Publication Data
A catalogue record for this book is available from the British Library

Library of Congress Cataloging-in-Publication Data
A catalog record has been requested for this book

ISBN 13: 978-0-367-58553-2 (pbk)
ISBN 13: 978-0-8153-7703-0 (hbk)

Typeset in Times New Roman
by Taylor & Francis Books

Contents

vi   *Contents*

Figures

Contributors

Andreas Benz is Senior Lecturer at the University of Augsburg, Germany. His research focuses on processes of socioeconomic transformation in rural societies of the Global South, with a particular interest in education, migration, social justice and the political ecology of human-environmental relations. His regional focus is on Pakistan, India and Cuba. He is author of *Education for Development in Northern Pakistan: Opportunities and Constraints for Rural Households* (Oxford University Press, 2014).

Jacob Bull is a Researcher in the Centre for Gender Research at Uppsala University, Sweden. He is editor of the *Crossroads* book series and member of the editorial collectives for *Humanimalia: A Journal of Human/Animal Interface Studies* and *Trace: Finnish Journal for Human–Animal Studies*. As a geographer interested in fish, ticks, cattle and bees, his work focuses on the role of animals in understandings of space, place and identity. He is the author of numerous book chapters and articles in geography journals, and has edited three collections: *Animal Places: Lively Cartographies of Human–Animal Relations* (Routledge, 2018), *Illdisciplined Gender: Engaging Questions of Nature/Culture and Transgressive Encounters* (Springer, 2016) and *Moving Animals: Essays on Direction, Velocity and Agency in Humanimal Encounters* (Uppsala University, 2011).

David Butz is Professor of Geography at Brock University, Canada, where he is a founding faculty affiliate of Brock's Social Justice Research Institute and the Social Justice and Equity Studies graduate program, and serves as editor-in-chief of *Studies in Social Justice*. He has been conducting ethnographic research in northern Pakistan since 1985 on topics relating to human transport labor, environmental governance, irrigation agriculture, mobilities, disaster recovery and development-related social change. Current research with Nancy Cook focuses on the differential mobility implications of a recently constructed jeep road linking Shimshal village to the Karakoram Highway, northern Pakistan's arterial roadway. His work has been published in edited collections, *The Sage Handbook of Qualitative*

Geography (Sage, 2010), *The International Encyclopedia of Geography* (Wiley, 2017) and numerous geographical and interdisciplinary journals.

Noel Cass is a Senior Research Associate in the Department of Sociology at Lancaster University, UK. His interests lie in mobility/transport and climate change policy, renewable energy technologies, nuclear waste disposal policy, fuel poverty, energy and the built environment and carbon capture and storage. He has extensive experience in public participation in policymaking, particularly conducting and analysing public engagement processes. Current work focuses on how energy demand is locked in through office building design processes. He has published widely on these topics in journals in geography, transportation research, architectural design, sociology and interdisciplinary science.

Georgine Clarsen is an Associate Professor and Discipline Leader of History at the University of Wollongong. Her research has focused on mobilities in settler colonial Australia as a distinctive constellation of raced and gendered practices. She is a founding editor of the journal *Transfers: Interdisciplinary Journal of Mobility Studies* and of the book series *Explorations in Mobility*, both published by Berghahn Press.

Nancy Cook is Associate Professor of Sociology at Brock University, Canada. Her research has focused on transcultural interactions between development workers from the Global North and local populations in Pakistan, focusing on their gendered, sexualized, racialized and imperial dimensions (*Gender, Identity and Imperialism: Western Women Development Workers in Pakistan*, Palgrave Macmillan, 2007). Recent research with David Butz entails two studies in the Gilgit-Baltistan region of northern Pakistan related to road infrastructure and associated mobilities. One focuses on the uneven mobility implications of a new link road that connects Shimshal village to the Karakoram Highway; the other investigates a range of demobilizations experienced in the Gojal region in the aftermath of a landslide disaster that destroyed a large section of its arterial roadway. This work is published in a number of book chapters and journals such as *Mobilities, Social and Cultural Geography, Contemporary South Asia* and *Studies in Social Justice.*

Anna R. Davies is Professor of Geography at Trinity College Dublin, Ireland. Her research interests lie at the intersection of environmental governance and sustainability, and she is currently examining the sustainability implications of new sociotechnical developments in urban food systems. A member of the Royal Irish Academy, Anna has advised the Irish government as an independent member of its National Economic and Social Council and National Climate Change Advisory Council. Anna is a board member of the European Roundtable on Sustainable Consumption and Production and a founding member of Future Earth's Knowledge Action Network on Systems of Sustainable Consumption and Production.

Andrew Gorman-Murray is Professor of Geography at Western Sydney University, Australia. His interests include gender, sexuality and space; urban transformations; mobilities and place-making; household dynamics and home/work interchange; emotional geographies and wellbeing; and disaster planning and emergency management. With Catherine Nash, he is exploring 'new' LGBT urban geographies. With Catherine Nash and Kath Browne, he is examining transnational resistance to LGBT rights. With David Bissell, he is investigating how mobile work is transforming Australian homes. He co-edited *Material Geographies of Household Sustainability* (Routledge, 2011), *Sexuality, Rurality and Geography* (Lexington, 2013), *Masculinities and Place* (Ashgate, 2014) and *Queering the Interior* (Bloomsbury, 2018). He is co-editor of the journal *Emotion, Space and Society*.

Suzan Ilcan is Professor of Sociology at the University of Waterloo, Canada. She is the author of *Longing in Belonging: The Cultural Politics of Settlement* (Praeger, 2002) and co-author of *Issues in Social Justice: Citizenship and Transnational Struggles* (Oxford University Press, 2013) and *Governing the Poor: Exercises of Poverty Reduction, Practices of Global Aid* (McGill-Queen's University Press, 2011). She is editor of *Mobilities, Knowledge, and Social Justice* (McGill-Queen's University Press, 2013). Her current Social Sciences and Humanities Research Council-funded research projects examine migration policies, bordering practices and humanitarian aid in the context of the displacement, precarity and mobility of refugees.

Ole B. Jensen is Professor of Urban Theory at the Department of Architecture, Design and Media Technology, Aalborg University. He has an educational background in political science, sociology and planning. He is deputy director, co-founder and board member at the Centre for Mobilities and Urban Studies (C-MUS). His main research interests are urban mobilities, mobilities design and networked technologies. He is the author of *Staging Mobilities* (Routledge, 2013) and *Designing Mobilities* (Aalborg University Press, 2014), editor of the four-volume collection *Mobilities* (Routledge, 2015) and author (with Ditte Bendix Lanng) of *Urban Mobilities Design: Urban Designs for Mobile Situations* (Routledge, 2017).

Fredrik Karlsson is a Senior Lecturer of Ethics at Dalarna University, Sweden. He has published works on a number of issues related to animal ethics and human-animal relations, including a monograph on animal rights theories and papers on anthropomorphism. His research presently focuses on morally connoted terminologies within ethological research, and on anthropomorphic and anthropocentric habits of interpretation.

Weiqiang Lin is Assistant Professor in the Department of Geography, National University of Singapore. His research interests include air transport, mobilities, logistics and the production of infrastructures. He also has a cognate interest in migration and transnationalism. In extension of his longstanding fascination with airspace production in Southeast Asia, his latest research focuses on the infrastructural underpinnings of the air logistics of food and cargo in Singapore and China (especially along new 'OBOR' corridors). His work has appeared in a number of book chapters, as well as international peer-reviewed journals, including *Environment and Planning A, Environment and Planning D, Geoforum, Journal of Transport Geography, Mobilities, Progress in Human Geography* and *Transactions of the Institute of British Geographers*. He is also a member of the editorial board of the journal *Mobilities*.

Heather Maguire teaches Communications in the Department of Applied Science and Environmental Technology at Algonquin College, and she works as a Research Assistant in the Department of Geography and Tourism Studies at Brock University. Her interests include mobilities and mobile technologies, and the social and cultural aspects of technological change. Her dissertation (2017) explores the relationship among mobility and information and communication technologies (ICTs) in life at sea for sailors on a cargo ship. Her work has appeared in *Area* (2018) and *NanoEthics* (2012).

Katharina Manderscheid is Professor of Sociology at Universität Hamburg, Germany. She has written on mobilities and social inequality, sustainable mobilities, mobile methods and urban sociology. Her current work focuses on car-free households and the future of automobility. She recently published on driverless cars ('From the auto-mobile to the driven subject: Discursive assertions of mobility futures', *Transfers, 8*(1), 24–43) and co-edited *The Mobilities Paradigm: Discourses and Ideologies* (Routledge, 2016), a book on discourse and mobilities. She is a board member of the European Cosmobilities Network.

Catherine J. Nash is Professor of Geography in the Department of Geography and Tourism Studies at Brock University, Canada. Her research focus is on sexuality, gender and urban places. Her current research interests include changing urban sexual and gendered landscapes in Toronto; international resistances to LGBT equalities in Canada, the UK and Australia; digital technologies and sexuality in everyday life and new LGBT mobilities. Her books include *Queer Methods and Methodologies* (Ashgate, 2010) with K. Browne and *Human Geography: People Place and Culture* (Canadian Edition) (Wiley, 2015) with E. Fouberg, A. Murphy and H. de Blij.

Denver V. Nixon is a Postdoctoral Research Associate in the Transport Studies Unit at the University of Oxford, UK. His current work as a member of the DePICT international collaborative research project explores community-led walking and cycling infrastructural initiatives in London and São Paulo to critically evaluate their potential contributions to *just* transitions in urban mobility. Before this, he studied how people's transportation mode shapes their understanding of their commute environments. More broadly, Denver is interested in how environmentally (un)sustainable and (un)just practices are formed and maintained through embodied sensory experience and particular social, cultural and material contexts.

Amy E. Ritterbusch is an Assistant Professor of Social Welfare at the UCLA Luskin School of Public Affairs. She has led social justice-oriented participatory action research initiatives with street-connected communities in Colombia for the last decade and recently in Uganda. Her recent work involves the documentation of human rights violations and forms of violence exerted against homeless citizens, sex workers, drug users and street-connected children and youth, and subsequent community-driven mobilizations to catalyze social justice outcomes within these communities. Her work has appeared in *Annals of the American Association of Geographers, Global Public Health* and *Alternate Routes: A Journal of Critical Social Research*.

Sharon R. Roseman is Professor of Anthropology and Associate Dean (Research) in the Faculty of Humanities and Social Sciences at the St. John's Campus of Memorial University of Newfoundland. She is interested in mobility histories in relation to migration, paid and unpaid labor, commuting, tourism and pilgrimage in Canada and Spain. She is author of *O Santiaguiño de Carreira: O rexurdimento dunha base rural no concello de Zas* and editor of four other books. She is co-director, with Elizabeth Yeoman, of a film on mobility justice activism for pedestrian rights in St. John's, titled *Honk If You Want Me Off the Road*.

Tim Schwanen is Associate Professor of Transport Studies and Director of the Transport Studies Unit at the University of Oxford, UK. He is interested in questions of radical change in mobility systems and in the relationships of justice, inequality and wellbeing with the everyday mobilities of people, goods and information between and within cities. He has published widely on these topics in journals in geography, urban studies, transportation research and interdisciplinary science.

Mimi Sheller is Professor of Sociology and founding Director of the Center for Mobilities Research and Policy at Drexel University in Philadelphia. She is founding co-editor of the journal *Mobilities* and past President of the International Association for the History of Transport, Traffic and Mobility. She has helped to establish the interdisciplinary field of mobilities research. She is author or co-editor of ten books, including most recently *Mobility Justice: The Politics of Movement in an Age of Extremes* (Verso, 2018), *Aluminum Dreams: The Making of Light Modernity* (MIT Press, 2014) and *Island Futures* (Duke University Press, forthcoming).

Gerard C. Wellman is Associate Professor of Public Administration at California State University, Stanislaus in Northern California. His research interests focus on social injustices in transportation planning, particularly with regard to public and active modes of transportation. His recent work has appeared in *Journal of Public Budgeting, Accounting & Financial Management, Public Works, Management, and Policy*, and *Public Administration Quarterly*.

Acknowledgments

This edited collection has its origins in a series of paper sessions collectively titled "Mobilities, Mobility Justice and Social Justice", which we organized for the March 2016 *Annual Meeting of the American Association of Geographers* in San Francisco. We extend our appreciation to the sessions' participants and attendees, whose stimulating presentations, questions and remarks motivated us to undertake this publishing project. Further energy and inspiration was derived from those who participated in the "Politics of Movement" sessions we organized for the *7th Nordic Geographers Meeting* held in Stockholm in June 2017. Thanks to those scholars as well.

We are grateful to the multidisciplinary array of authors who accepted the difficult assignment of contributing to the development of a nascent concept – mobility justice – on a tight timeline and through brief, accessible chapters. To a person they responded diligently, rigorously and with good humor to our many no doubt irritating requests for revisions and refinement. We appreciate their cooperation and their commitment to the overarching project of conceiving mobility and social justice in relation to one another.

Portions of Chapter 6 were previously published in different form in the journal article listed below, and are used here with the authors' permission according to Creative Commons License BY-ND-NC:

Butz, D. & Cook, N. (2017). 'The Epistemological and Ethical Value of Autophotography for Mobilities Research in Transcultural Contexts'. *Studies in Social Justice, 11*(2), 238–274.

Part I
Introducing mobility justice

1 Moving toward mobility justice

Nancy Cook and David Butz

Introduction

Mobility is a marker and maker of 21st-century social life. Interconnected and intensifying flows of people, animals, goods, information and waste, and the infrastructures and technologies that facilitate them, fundamentally shape everyday life, regional and national processes and global and planetary orders. These mobilities can be significant sources of social advantage; for example, by enabling people to access economic opportunities and goods, develop identities and social networks, participate in civic life and pursue personal ambitions. Mobility is a resource for (non)human agency and the performance of daily life. But mobilities also have differential and troubling consequences, as attested by climate change and those marginalized social groups immobilized in the slow lanes of life or forced into the express line.

What productive theoretical, empirical and political outcomes can be achieved by considering the array of mobilities that characterize and produce contemporary socio-material life and their differential consequences for human and more-than-human existence through the lens of social justice? Our goal in this collection is to explore the relationship between mobilities and justice in a variety of social and spatial contexts, and at a number of scales, by drawing into conversation 'mobilities paradigm' scholarship and social justice theorizing, and from that intersection to develop the concept of 'mobility justice'.

This dialogue has important implications for both bodies of literature. Mobilities research focuses on the 'politics of mobility' or the socio-spatial inequalities that are (re)produced as a result of differential access to or effects of various kinds of mobility. How might social justice concepts strengthen its analyses of mobility as socially structured movement in fields of power? What new justice-related questions emerge by thinking about uneven mobilities in a social justice frame? On the other hand, social justice scholarship has yet to engage deeply with the dynamics and effects of a contemporary world constituted through spatially diverse networks of people, objects and infrastructures. What happens to its analytical object – the social – when it is understood in terms of the "complex relationality of places and people connected through movement", often across wide spatial expanses (Sheller & Urry, 2006, p. 213)? The ontological politic driving this collection suggests that the mobile ontology developed by mobility scholars can strengthen theories of social justice, making understandings of justice relevant to 21st-century social relations. And what does theorizing 'mobility justice' achieve for both bodies of scholarship? The concept draws together the many scales, types and facets of mobility with multiple approaches to social justice to reimagine justice as a mobile concept appropriate for analyzing and effecting the ethicality of contemporary mobile life.

The politics of mobility

The interdisciplinary 'mobilities paradigm' was introduced about a decade ago as a response to social science's characteristically sedentarist approach:

> The mobilities paradigm challenges the ways in which much social science research has been "a-mobile". Even while it has increasingly introduced spatial analysis, the social sciences have still failed to examine how the spatialities of social life presuppose (and frequently involve conflict over) both the actual and imagined movement of people from place to place, person to person, event to event.
>
> (Sheller & Urry, 2006, p. 208)

Paradigm founders argue that accounting for the mobilities that connect people and places in spatially dispersed networks requires social scientists to reimagine objects of inquiry.

Mobility scholars achieve this reorientation by foregrounding mobility as a foundational social relation (Hannam, Sheller & Urry, 2006) and by employing ontogenetic notions of reality to argue that movement and flows continually (re)shape social and material entities conventionally understood as stable, bounded or sedentary (Amin & Thrift, 2002; Latour, 2005; Lefebvre, 1974). John Urry (2008, p. 13), for instance, demonstrates that social relations are "never only fixed or located in place. ... There are many circulating entities that bring about relationality within and between societies at multiple and varied distances". Mobilities, broadly conceived, are crucial to circulations that enable social life to be networked across space and render borders porous and unstable. The mobility agenda intensifies analytic attention to these "horizontal fluidities" (Urry, 2000, p. 2) that mark our times, the dynamism in systems of movement that accentuate routes rather than roots (Clifford, 1997).

As intensified circulation becomes a fundamental social organizing principle, the ability to move – or to choose to stay put – is socially advantageous, making mobility capital (Kaufmann, Bergman & Joye, 2004) and network capital (Elliott & Urry, 2010) valuable social resources that, like other forms of capital, are unevenly accessed by or have differential effects for particular social groups, thereby producing new socio-spatial opportunities and inequalities. In other words, as social relations are spatially distantiated, mobilities become spatial mechanisms through which social inequalities are (re)produced.

Reduced or forced mobility, for example, hinders people's access to social services, networks and opportunities, and their ability to maintain "meetingness", producing social exclusion (Elliott & Urry, 2010, p. 59). Mobilities, therefore, are socially differentiated and unevenly experienced, as they are co-constituted in relation to hierarchies of gender (Cook & Butz, 2017; Cresswell, 1999), race (Nicholson & Sheller, 2016), sexuality (Montegary & White, 2015), disability (Goggin, 2016) and empire (Skwiot, 2009). They operate in fields of power as socially structured movement to create and reinforce social difference.

Attending to "the ways in which mobilities are both productive of [inequitable] social relations and produced by them" (Cresswell, 2010, p. 21) leads researchers to track

> the power of discourses, practices and infrastructures of mobility in creating the effects of both movement and stasis. ... [M]obilities research interrogates who and what is demobilized and remobilized across many different scales, and in what situations mobility or immobility might be desired options, coerced or paradoxically interconnected.
>
> (Sheller, 2011a, p. 2; see also Cook & Butz, 2015)

It also investigates the processes through which movement and stasis are governed to produce inequitable socio-spatial arrangements. The driving question for mobility scholars thus becomes: How are particular social actors situated in relation to various (im)mobilities, and to what political effects? Our collection extends this question by asking: How can theorizing these political effects in terms of justice strengthen mobility analyses? What conceptual tools from the vast social justice literature may be useful for such an endeavor? In the following section we negotiate an eclectic, and by no means thorough, way through some prominent social justice theory to provide tentative suggestions, which are explored in later chapters.

Theorizing social justice

Social justice theorizing largely has been the purview of political philosophy. Justice scholars have traditionally focused on "transcendental institutionalism" (Sen, 2009), which has its roots in the collective works of Hobbes, Kant, Locke, Rawls and Rousseau. These theories of justice address questions about the nature of perfectly just institutions and rules; their answers provide *a priori* principles and schemas for realizing institutional perfection, from which just social relations and ideal social arrangements ostensibly unfold. Hobbes and Rawls emphasize the crucial role played in this exercise of justice by the sovereign state as the political entity that establishes perfect institutions through which principles of justice are applied. Those principles relate primarily to the just distribution of social benefits and burdens. As Rawls (1971, p. 9) argues, "a conception of justice [provides] in the first instance a standard whereby the distributive aspects of the basic structure of society are to be assessed". This entanglement of institutionally derived justice, the nation-state and distribution became the foundation of social justice theorizing, even into the late 20th century.

Miller (1999, p. 250), for example, contends that

> to achieve social justice we must have a political community in which citizens are treated as equals across the board, in which public policy is geared toward meeting the intrinsic needs of every member and in which the economy is framed and constrained so people receive in correspondence with their deserts.

This claim rests on three premises. First, justice requires a bounded society that constitutes the frame of distribution. Why? Consistent with transcendental institutionalist thinking, Miller (1999, pp. 5–6) argues that we need a delineated universe of distribution within which fairness can be judged, and the state is the organic choice because its members share values and understandings of need, it generates trust among members, and it has the means to invoke constraint. Second, states are positioned to devise just social institutions and policy/rules that shape citizens' advantages; they are effective top-down mechanisms of (re)distribution. Third, states can alter unjust institutional structures.

Three challenges to this justice-state-distribution nexus have recently developed within the realm of social justice theorizing. The first relates to institutional design as the mechanism for achieving justice. Sen (2009) argues for a more grounded approach that diagnoses redressable injustices in everyday life. This 'realization-focused approach', in contrast to the established 'arrangement-focused approach', asks how lived injustice can be identified and removed, rather than how to devise perfectly just institutions as a way to advance justice. Young (1990) similarly focuses on everyday experiences of injustice voiced by American social movements. She argues that these experiences disrupt political philosophy's dominant understanding of justice by advancing alternative conceptualizations. While institutional structure retains significance in her work as a mechanism that produces injustice, Young eschews *a priori* institutional principles in favor of grounded claims of lived injustice as the starting point for theorizing about social justice. Together, these realization approaches orient us to injustices grounded in the lived realities of different social groups.

Second, the synonymy of social justice and distribution has been questioned. Sen's (1993) capability approach is a noteworthy example of how dimensions of justice have expanded beyond distributional bounds. He contends that a theory of justice should provide tools for assessing an individual's overall advantage, which is effectively judged by their capability to do the things they value. Justice, therefore, entails the equality of capabilities, freedoms that allow individuals (and social groups) to live lives they value, thereby producing social opportunity and wellbeing. Young (1990) expands facets of justice further by critiquing theories that reduce the justice 'community' to an undifferentiated unity, which excludes considerations of 'social group' differences. Her critique of the ontology of sameness leads to a theory about justice that foregrounds injustices experienced and articulated by particular social groups, including domination (constraints on self-determination imposed by having to follow rules established by others) and oppression (constraints on self-development, or on the ability to develop and exercise capacities, and to express experiences and be intelligibly heard). Narrowly focusing on distribution renders invisible these injustices and ignores the institutional contexts within which distribution occurs. Institutional contexts include "any structures or practices, the rules and norms that guide them and the language and symbols that mediate social interactions within them, in institutions of state, family and civil society, as well as the workplace" (Young, 1990, p. 22). These components "condition people's ability to participate in determining their actions and their ability to develop and exercise their capacities" (Young, 1990, p. 22). Her concept of justice, therefore, foregrounds institutional constraints on self-determination and self-development, as well as mechanisms and contexts of distribution.

Third, theorists are grappling with the globalized context of contemporary life and its implications for understanding and achieving justice. At one end of the globalization-justice continuum sit scholars who view the conceptual pairing as oxymoronic. Miller (1999, p. 6), for example, argues that the three premises outlined above define the circumstances of the transcendental institutionalist approach such that

> if we do not inhabit bounded societies, or if people's shares of goods and bads do not depend on social institutions, or if there is no agency capable of regulating that basic structure, then we no longer live in a world in which the idea of social justice has any purchase.

'Global justice', he argues, cannot be understood on the model of 'social justice' because the key premises of justice do not hold at the global level. Miller thinks there remains a limited place for the idea of social justice through the roles states can play in regulating global economic activity, strengthening welfare mechanisms and reinforcing their authority in deliberations about the global economy. Thomas Nagel (2005) is even less optimistic about social justice prospects in the globalizing world. His position is that 'global justice' demands a sovereign global state; without such a political entity that can satisfy the institutional demands for a just world at the global level, the question of global justice is moot. For scholars working within transcendental institutionalist parameters, neither social nor global justice seems feasible in a world of intensified transnational flows and networks of connection.

Capeheart and Milovanovic (2007, p. 80) disagree, arguing that "principles of social justice should not fly in the face of globalization. Rather, we should reconsider our ideas of social justice to ensure they are compatible with new global realities". Two advocates of this reorientation project are noteworthy (see also Young, 2002, on justice as inclusive democracy in a globalizing world). Sen (2009) develops a global politics of justice by focusing on 'second-generation' human rights – social and economic rights beyond basic freedoms – that gained global currency in the late 20th century. He contends that as injustices like poverty are framed globally as rights violations, states are pressured to reform economic policy. Because valuable capabilities are the basis for human rights and social justice, global justice will involve states and transnational institutions ensuring that globalization processes do not hinder the rights and capabilities of anyone in the global population. Global justice becomes the extended integration of adequate levels of agency and basic capabilities, achieved through the institutional promotion and regulation of second-generation rights.

Nancy Fraser (2009, p. 16) takes a different tack by defining justice as participation parity: "justice requires social arrangements that permit all to participate as peers in social life. Overcoming injustice means dismantling institutional obstacles that prevent some people from participating on a par with others, as full partners in social interaction". Informed by earlier debates with Young, Fraser contends that unjust economic structures deny resources for interaction (a matter of distributive justice), while hierarchies of cultural value (an issue of recognition) deny some social groups standing for participation. While these injustices are conventionally imagined within the bounds of domestic political space and among fellow citizens, Fraser notes that contemporary claims for redistribution seldom invoke only national economies, and claims for recognition usually look beyond the territorial state. To accommodate these realities, social justice theory, she argues, needs conceptual resources appropriate to a global frame of justice in terms of participation parity.

She develops them by introducing a new dimension of justice – political representation – that addresses the processes by which globalization limits the boundaries and participation of the justice community. Individual states cannot control unjust global forces, but people's claims about these injustices can only be articulated in domestic political spaces (a 'misframing' of political space). This has troubling consequences: (a) states cannot temper transnational powers, so the latter are shielded from justice; (b) offshore noncitizens who are similarly affected by unjust processes are sidelined in these national debates, excluding them from the justice community (an act of political 'misrepresentation'); and (c) states become powerful tools of injustice. To solve these problems, Fraser (2009, p. 24) calls for a transformative politics of framing, which involves applying "the all-affected principle": "all those affected by a given social structure or institution have moral standing as subjects of justice in relation to it". Justice would no longer be framed within the bounds of nation-states, and subjects of justice would constitute a community based on structural co-imbrication rather than propinquity as a governing mechanism of social interaction. Fraser's reframing of justice offers a globalized theory of justice and a new dimension of justice appropriate to an interconnected world.

So, what conceptual resources does this slice of social justice scholarship provide scholars interested in analyzing striated mobilities and developing the concept of mobility justice? Justice entails thinking critically about the many scales of justice and their complex interactions, from individual bodies and social groups to national and transnational frames. Attention should be paid to the institutional organization, control and governance of justice, but in relation to everyday experiences of injustice grounded in the lived realities of different social groups. And this scholarship delineates many aspects of justice as useful analytical tools: rights, freedoms, capabilities, distribution, domination, oppression (exploitation, marginalization, powerlessness, cultural imperialism, violence), participation, accessibility to (a)venues of and information about participation (producing social inclusion or exclusion), decision-making processes, understanding and recognition gaps and the vulnerabilities these deficits create. These dimensions of justice can be integrated into a typology of justice useful for considering the intersection of mobility and justice (see Capeheart & Milovanovic, 2007, and Sabbagh & Schmitt, 2016, for comprehensive overviews):

- distributive justice
- procedural justice
- deliberative justice
- restorative justice
- epistemic justice
- environmental/ecological justice
- retributive justice
- recognition.

While theories about social justice have much to offer mobility paradigm scholars, they have two main limitations (from a mobilities perspective) that prevent them from fully engaging with justice dimensions of 21st-century social relations. First, even though some engage with the challenges globalization presents to traditional understandings of justice, they have not moved beyond a 'societal' framing toward a mobile ontology so as to remain theoretically, empirically and politically relevant. Second, they do not consider justice in relation to more-than-human social relations (but for work on animals and justice see Talbot, Price & Brosnan, 2016), including the material objects, information and images, and the mobility platforms and technologies, through which contemporary social relations are constituted. In the following section we develop these critiques as part of our ontological politics as a way to demonstrate the possibilities of a mobile ontology for social justice scholarship.

Mobilizing social justice theory

These limitations point to the 'sedentarist' character of social justice theorizing in two ways: it does not appreciate that social relations are produced through (im)mobilities or that mobility constitutes an exercise of power as it shapes inequitable social relations. We argued these claims earlier in the chapter (as do upcoming chapters), so here we review what we mean by a sedentarist ontology and how explanations premised on it lack interpretive power in contemporary times, we demonstrate how sedentarism organizes theories about justice, and we suggest an alternative ontology appropriate to an (un)just world constituted through (im)mobilities.

Mobilities researchers have detailed scholars' ontological predilection to hold space still, by normalizing place, rootedness, dwelling, stability and belonging as a social 'steady state', and to understand "'terrains' as spatially fixed geographical containers for social processes. ... [Sedentarism] locates bounded and authentic places or regions or nations as the fundamental basis of human identity and experience. ... It rests on forms of territorial nationalism" (Sheller & Urry, 2006, pp. 208–209). A subordinate "nomadic metaphysics" operates in conjunction with sedentarism, as when mobility is associated with freedom and progress (Cresswell, 2006). Sedentarist ideas undergird dominant understandings of 'society', 'social structures', 'citizenship' and 'governance' in which static social relations are ostensibly produced, governed and bounded by the nation-state. They simultaneously render migration, displacement and border-crossing as threatening (think of dominant representations of tramps, nomads, immigrants and refugees). In this ontological frame, dwelling is morally and ideologically privileged; mobility is suspect beyond simply moving between spatial points.

Considering the research that demonstrates how flows and networks of connection among people, objects, information and material infrastructures constitute contemporary places and social relations, does 'society' as a territorial unit bounded by the nation still exist? Are identities and cultures created only through place embeddedness? And are belonging and governance merely regional or national practices and preoccupations? John Urry's (2000, p. 2) answer to these questions rests in his observations that a range of global mobilities have undermined territorially bounded social structures, launching a 'post-societal' reality: "material transformations are remaking the 'social', especially those diverse mobilities that, through multiple senses, imaginative travel, movements of images and information, virtuality and physical movement, are materially reconstructing the 'social as society' into the 'social as mobility'". Societies of flow are interdependent, edgeless, fluid entities shaped by global processes, understandable only through an interrelationship produced through chains of connection. Living beyond the nation (mostly obviously through the Internet, consumption, global money markets and world travel), we develop post-societal identities, cultures, governance structures and senses of (in)justice. In this context, "all states can do is regulate these diverse mobilities that transform nation state and civil society" (Urry, 2000, p. 195).

If we are convinced by the argument that contemporary social relations are configured through mobilities, then we appreciate that global networks and flows (re)structure social inequalities. Manderscheid (2009, p. 11) calls for the unbounding of injustice by arguing that "the territorial spatial frame of the nation-state ... produces and structures fewer and fewer dimensions of social inequality, and, therefore, national borders are, in many dimensions of social inequality, no longer the borders of relevant power relations". Current environmental, economic and political injustices operate at the global level, as do demands for human and migration rights and calls for a global citizenship aligned with an interest in global wellbeing. A sedentarist ontology provides little interpretive space for describing these actually occurring social processes, leaving its analytic usefulness open to question.

In postulating static social relations, sedentarism also neglects materiality in terms of the flows of materials and mobility infrastructures and technologies through which social life is constituted across space, further constraining its analytic potential. As mobility scholars describe, the material stuff of life – waste, consumer goods, mobile phones, automobiles, computer software – is in motion, structuring social realities: "What we call the social is materially heterogeneous: talk, bodies, texts, machines, architectures, all of these and many more are implicated in and perform the social" (Law, 1994, p. 2). Systems of materialities organize 'post-societal' life, but they also constitute the physical means for movement. Roads, airports, rail lines, ports and fiber-optic cables – these infrastructural moorings configure and enable physical and virtual mobilities, thereby constituting fundamental aspects of mobility systems (Hannam, Sheller & Urry, 2006; Sheller & Urry, 2006). The 'social as mobility' is realized, in large part, through material infrastructure (Cook & Butz, 2018).

And so is justice. Landscape and urban planning scholars (e.g., Jackson, 1984; Martens, 2012, 2016) demonstrate that transportation infrastructure has contradictory justice implications. On one hand, it can extend a person's mobility capabilities and options and, consequently, their social engagements. Less sanguinely, roads, for example, also articulate relations of social inequality by entrenching the social marginalization of those who lack access to an automobile (Featherstone, Thrift & Urry, 2005) or live in rural regions (Farrington & Farrington, 2005). Stranded, the spatially marginalized are less agential at some scales. Studies of highways and other urban transit systems illustrate the structuring power of infrastructure; they are mobility platforms that enhance or compromise justice depending on how they influence different social groups' ability to participate in movement.

How does a sedentarist ontology organize theories about social justice, beyond excluding considerations of material infrastructure as co-constitutive of (in)justice? As we mention above, they retain the 'social as society' as an analytic object by neglecting the global networks and flows that produce places and social inequalities, as evidenced most clearly in transcendental institutionalism. Sen's (2009, p. 232) capability approach also presupposes bounded societies: it

> is a general approach, focusing on information on individual advantages, judged in terms of opportunity rather than a specific "design" for how a society should be organized. ... In judging the aggregate progress of a society, [it] would certainly draw attention to the huge significance of the expansion of human capabilities to all members of a society.

And his notion of global justice involves independent states regulating exogenous forces of globalization to ensure citizens' rights and capabilities, as if those states and citizens were not always already shaped by the global flows that require regulation. Fraser (2009) similarly recognizes structures that 'exceed' domestic political boundaries, but she also does not incorporate stretched-out social relations and mobilities into the structural framework that governs social interaction. Acknowledging that some inequitable structures (e.g., financial markets, information networks) operate at the global scale does not fully appreciate the 'social as mobility', leaving justice an immobile concept in a mobile world of injustices.

Second, established justice frameworks do not consider how distributions of and accessibility to social 'goods' (e.g., capital, knowledge) is predicated on mobility, which in turn requires particular arrangements of those goods. If we understand 'access' as "the ability to negotiate space and time to accomplish practices and maintain relations that people take to be necessary for normal social participation" (Cass, Shove & Urry, 2005, p. 543), then access requires networks and mobilities. Without understanding social life as networked across space, social justice scholarship develops anemic notions of 'access' and 'participation'. For instance, Fraser's participation parity requires equitable mobility systems; otherwise, people cannot participate on a par with others as full partners in social interaction. Immobilization is an obstacle for participation, as is forced or involuntary movement.

Third, social justice theorists overlook the mobility dimensions of social exclusion, which is often the outcome of "some combination of distance, inadequate transport and limited ways of communicating" (Cass, Shove & Urry, 2005, p. 539). According to Kenyon, Lyons and Rafferty (2002, pp. 210–211), social exclusion is

> the process by which people are prevented from participating in the economic, political and social life of the community because of reduced accessibility to opportunities, services and social networks, due in whole or in part to insufficient mobility in a society and environment built around the assumption of high mobility.

As aspects of social justice, social inclusion, accessibility and participation require voluntary (im)mobility and the ability to form and maintain a range of distantiated social networks. Enabling 'meetingness' is a key justice dimension in post-societal contexts.

In summary, theories of social justice are predicated on a sedentarist ontology that inadequately captures a world constituted through multi-dimensional and multi-scalar mobilities enabled by material infrastructural moorings. They have yet to consider the horizontal (im)mobilities that shape 21st-century social life and its inequalities. We need a mobile ontology for a mobile world which can move social justice theorizing into the 'post-societal' present so as to remain theoretically, empirically and politically relevant. Achieving this approach involves rethinking theories of justice within what Mimi Sheller (Chapter 2) calls a "kinopolitical mobile ontology", in which movements regulated by power are understood as a "foundational condition of being, space, subjects [and resistance]" (p. 24), thereby rendering justice a more mobile concept. Mobility scholars have taken up this challenge by developing the notion of 'mobility justice'.

Theorizing mobility justice

Movement toward 'mobility justice' has involved several conceptual steps within the politics of mobilities literature. It begins with work invoking concepts like freedom, rights and exclusion, without explicitly foregrounding them as dimensions of social justice, to analyze how power relations shape the control, practice and representation of movement and inequitable movement patterns (e.g., Grieco & Hine, 2008; Mountz, 2011; Sager, 2006). Next, more explicit connections between social justice and mobilities are made in research on the justice-mobility-knowledge nexus, mobility as a capability and 'transport justice'. Ilcan's (2013, p. 11) edited collection entwines social justice and mobilities theorizing by demonstrating "how some forms of mobility generate specific types of knowledge, how knowledge shapes patterns of mobility and how this mobility-knowledge nexus creates new challenges to and opportunities for the realization of social justice" in its distributive form. Kronlid (2008) and Uteng (2006) relate mobility and social justice by arguing that justice entails enhancing capabilities, of which mobility is one. Because movement options are intrinsic to human wellbeing and flourishing, they argue, mobility is an independent capability in the Sensian sense.

The notion of 'transport justice' (Golub & Martens, 2014; Martens, 2012, 2016; Pereira, Schwanen & Banister, 2017) is the most obvious affiliate to the concept of mobility justice. Martens and his colleagues theorize 'transport fairness' in relation to the distribution of transport benefits and people's accessibility to them as a means of connecting them to spatially separated activities and opportunities (fairness as the just distribution of possibilities for movement). They argue that people's ability to access and appropriate potential transport mobility options is constrained by many factors, including residence location, knowledge of the transport system, gender, household composition, access to transport funds and physical ability. Justice, therefore, lies in a 'fair' transport system designed with these constraints in mind to provide "a sufficient level of accessibility to all under most circumstances" (Martens, 2016, p. 151). Pereira, Schwanen and Banister (2017) employ a capabilities approach similarly to emphasize fairness in the distribution of (and accessibility to) mobility to prevent social exclusion (see also Ureta, 2008) and of the risks and benefits of mobility infrastructures. Extending the notion of fairness, they also call for greater attention to procedural justice – just decision-making and participatory processes in transport planning – and transport mobility rights and entitlements that are often unrecognized and unenforced.

Finally, several authors have begun explicitly theorizing 'mobility justice' in ways that frame justice beyond the realm of transportation and urban accessibility. Sheller (2011, 2012, 2015) coined the term, which has been developed in disparate ways (Cook & Butz, 2016; Mullen & Marsden, 2016; Vukov, 2015), as authors discuss in upcoming chapters. But there is yet no common agreement on its meaning and parameters. It was not our intention at the outset of this publishing project to impose on contributors a particular and singular *a priori* understanding of mobility justice. Rather, we provided a platform for an interdisciplinary and spatially diverse group of mobility and social justice scholars to think through more comprehensive understandings of the concept, framed by a kinopolitical mobile ontology and drawing on different mobility contexts, scales, mobile agents and dimensions of justice that are helpful in developing holistic, multifaceted and intersectional analyses of mobility-related injustices. Sheller (Chapter 2) provides a useful framework for this thinking project by suggesting that mobility justice is an overarching concept for considering "how power and inequality inform the governance and control of movement, shaping the patterns of unequal mobility and immobility in the circulation of people, resources and information" (p. 23).

By employing a mobile ontology, contributors render intelligible a range of mobility-related (in)justices that entail embodied relations of race, gender, class, age and sexuality, involve governing institutions/processes and material infrastructures, and operate at multiple (and sometimes entangled) scales including the body, household, community, city, region, nation and globe. These include:

- transport justice and urban accessibility
- environmental justice and sustainability
- spatial justice (exclusions from public space, the right to the city, the right to dwell)
- migrant justice
- infrastructural justice
- epistemic justice
- recognition and sovereignty
- more-than-human justice.

Following Chapter 2, in which Mimi Sheller develops a holistic approach to theorizing mobility justice and a comprehensive schema for analyzing the intersectional aspects of mobility inequities, these discussions are developed in 15 chapters, organized into four sections.

Justice and mobility governance

In this section, contributors focus on two ways that mobility-related injustices are produced in relation to acts of governance. First, the governance *of* mobility involves institutional control of mobility and movement systems at several scales. This (re)designing of mobility milieus often produces unequal patterns of (im)mobility. In Chapter 3, Weiqiang Lin examines the global governance apparatuses that regulate air travel to produce variegated mobilities among nations, thereby frustrating equal aeromobile opportunities and the mobility rights of populations in the Global South. Under circumstances of aeromobility exclusion, mobility justice entails altered governance mechanisms that fairly redistribute aeromobility by guaranteeing everyone the ability to exercise rights to it.

In Chapter 4, Suzan Ilcan focuses on governance of Syrian refugees fleeing civil war through bordering practices that attempt to curtail their mobility. She details refugees' encounters with and resistance to unjust 'state mobilities', as well as the demands for migrant and mobility justice that develop through these encounters, which entail freedom of movement, safety, protection, human rights and citizenship. In this migration and border context, mobility justice involves developing national and international systems of care, rights and citizenship that advantage migrant populations forcibly escaping war and persecution.

In Chapter 5, Gerard C. Wellman discusses governance of mobility through transportation planning in the United States. He argues that planners usually approach mobility provision as a policy problem involving safety, cost and adequate infrastructure without considering the power relations exercised through decision-making processes. To do so, he examines how planning often reduces access to mobility-related goods for marginalized groups, thereby serving as a tool of exploitation that preserves existing class structures and denies poor people the freedom to determine how, where and when they move. In this context, mobility justice involves eliminating exploitation in the field of transport planning, which begins with planners' more fulsome evaluation of its unjust everyday social consequences.

Second, governance *through* mobility – or 'governmobility' (Bærenholdt, 2013) – positions mobility as a governing mechanism that entails mobility practices (e.g., tourism, migration) designing and regulating social relations, self-government processes and the conduct of others as exercises of power and the production of (un)just social arrangements. Mobilities, in short, are fundamental to fashioning and governing social relations. In Chapter 6, David Butz and Nancy Cook use their research context in northern Pakistan to illustrate how mobilities research that employs 'go-along mobile methods' in particular contexts is a powerful mechanism of governance that regulates the conduct of research participants in ways that produce epistemic injustice. They argue that methodological instantiations of epistemic injustice in mobilities research constitute mobility injustices. Mobility justice, therefore, involves researchers carefully considering the ethical implications of their methodological decisions.

Justice and mobility infrastructures

As we have argued, material infrastructure configures mobilities and is co-constitutive of injustice through its ability to differentially enhance or compromise the circulation of people, information and objects. Mobility justice pertains to infrastructure, for example, in unequal distributions of auto-mobility, public transport, cycling access and mobility infrastructures' risks and benefits. In Chapter 7, Noel Cass and Katharina Manderscheid critique the social norm of mobility as a basic need for the 'good life' and associated compulsions to be auto-mobile that create environmental and justice problems. They argue that the automobility system is unjust and unsustainable because it differentially imposes environmental harms and curtails freedom from compulsion and the autonomy not to move and pursue alternative ends. The model they develop to replace automobility is "autonomobility", which foregrounds the freedom from compulsion to move or to stay as the normative basis necessary for realizing mobility justice in the context of resource scarcity. This model proposes collective forms of movement for just social relations, situating human flourishing in relation to environmental limitations and distributional equity.

In Chapter 8, Ole B. Jensen discusses the "dark design" of mobility infrastructures, demonstrating how design ideas materialize in physical objects that regulate human movement in socially exclusionary ways. The exercise of power manifests in the materialization of mobility artifacts. Concentrating on the design of mundane city squares and public spaces, and the infrastructural developments facilitating the Holocaust, he shows how spatial confinement and social exclusion of "unwanted subjects" is achieved through rail lines and "bum-proof" urban furniture, that forcibly move them along and relegate them to particular, often deadly places. For Jensen, mobility justice and social inclusion are matters of material as well as social design.

In Chapter 9, Denver V. Nixon and Tim Schwanen focus on community initiatives in London and São Paulo that provide or enhance cycling and walking infrastructure. Starting from insights inductively derived from interviews with initiative leaders, staff and users, they analyze the multi-scalar, emergent and extemporaneous conceptions of justice that animate these infrastructural development practices, which they show are more spatially and temporally complex than academic understandings of transport and mobility justice. Drawing from Derrida and Daoism, they provide conceptual resources for understanding mobility justice in ways that resonate with these everyday notions of mobility-related justice.

In Chapter 10, Sharon R. Roseman analyzes the demand for mobility justice by residents of Bell Island, Canada, who, in the face of lost mining jobs at home, undertake a daily commute by ferry to access new city-based jobs on the large island of Newfoundland. Access to affordable and dependable aquamobility infrastructure is key to this demand, which enables islanders' 'right to the city' and 'right to dwell' on their island home and defends ferries as counter-mobilities vis-à-vis the unjust hegemony of land- and air-based travel.

Justice and biomobilities

Mobility-related injustices also pertain to micro-scale embodied relations of race, gender, age, disability, class and sexuality that, as exercises of power, shape uneven freedom of bodily movement. Mobility justice therefore involves the mobile processes through which differentiated subjects are constituted and resist injustice. In Chapter 11, Georgine Clarsen analyzes *Black As* – an Indigenous television series filmed in Australia – as an Indigenous expression of self-determined mobility. Such enactments are politicized in the context of oppressive Australian settler-colonial relations that impose uneven mobilities on Indigenous subjects. The characters move outside settler mobility regimes, thereby reanimating ancient cultural practices and refusing the political forces that have constrained Indigenous mobilities and sovereignty. Mobility justice is enacted through this embodied movement-based resistance and assertion of ongoing sovereignty.

In Chapter 12, Amy E. Ritterbusch focuses on a police raid targeting homeless people in Bogotá, Colombia, which involved forcibly and violently displacing them as part of a state-sanctioned urban renewal initiative. She analyzes homeless citizens' movement patterns and lived experiences of forced movement during this intervention, as well as media representations of it. The chapter examines a case of state domination through forced mobility that violates homeless citizens' right to dwell and occupy public space in the city.

In Chapter 13, Catherine J. Nash, Heather Maguire and Andrew Gorman-Murray consider new queer mobilities and reconstituted LGBTQ inner-city neighborhoods in Canada and Australia – outcomes of recent social justice achievements – in relation to the co-constituted processes of just public urban spaces and mobility justice. Using a framework delineating five aspects of justice, they evaluate how changes in queer mobilities, neighborhoods and access to public space materialize in increasingly just urban spaces and mobilities.

In Chapter 14, Andreas Benz argues that migration has driven social transformation in the Gojal region of northern Pakistan by improving villagers' access to outside resources and opportunities through translocal livelihood strategies and the social capital of spatially diverse kinship networks. However, his case study of two Gojali villages shows that growing economic disparity in the region is attributable to mobility injustice derived from unequal network access by gender and household positioning. He concludes that achieving mobility justice between and within households requires a more equitable distribution of mobility capital and access to mobility options.

Justice and more-than-human mobilities

Although mobilities scholarship has paid extensive attention to the movement of commodities, images and information, mobility and transport justice research has been dominated by justice issues related to human mobilities. There is ample scope therefore to consider justice implications in relation to other movements, including that of food and animals.

In Chapter 15, Fredrik Karlsson's philosophical treatise on mobility and justice draws on non-representational theory, which frames agency as a human and more-than-human phenomenon, to argue that the capability of mobility for all beings is the foundation of justice as a virtue. He expands the scope of mobility justice by exploring the relationships among autonomy, agency, perception, intentional action and turbulence to conclude that moral agency and mobility are mutually constitutive processes that include animal agents, which are worthy of respect due to their mobile existence and capacity for just acts.

In Chapter 16, Jacob Bull reinforces the argument that mobility justice is a more-than-human phenomenon by describing how human, tick and other animal movements in Sweden, and the representations of and affects attached to their mobilities, produce pathogenic landscapes and circulations. Vulnerabilities and risks are produced through pathologized movements which subject these beings to unjust governance as killable or valuable, with implications for their ability to live and die well together.

In Chapter 17, Anna R. Davies examines the relationship between food mobilities and social justice in the context of surplus food redistribution processes between businesses and charities in the United Kingdom. Focusing on one redistribution initiative, she analyzes how various actors represent their institutional goals, practices and impacts, from which she details the logistics of surplus food redistribution for human consumption. She also evaluates the benefits and drawbacks of redistribution initiatives for reducing food waste and food injustice and concludes that attending to distributional food justice in terms of access is insufficient to realize food mobility justice more broadly.

Altogether, contributions to this collection offer novel, productive and exciting insights into the relationship between mobility paradigm thinking and social justice scholarship, and into the concept of mobility justice that develops from considering them together in a kinopolitical mobile ontology. Authors theorize mobility justice from a range of disciplinary locations and analytic areas, in its multidimensional forms and multi-scalar and intersectional modes of operation. We hope that the chapters develop understandings of differential mobilities and social justice in ways that stimulate further analyses, strengthen the justice aspect of politics of mobility thinking and mobilize social justice theorizing fully into the post-societal present to tackle a range of mobility-related injustices.

References

Amin, A. & Thrift, N. (2002). *Cities: Reimagining the urban.* Cambridge: Polity Press.

Bærenholdt, J. (2013). 'Governmobility: The powers of mobility'. *Mobilities*, 8(1), 20–34.

Capeheart, L. & Milovanovic, D. (2007). *Social justice: Theories, issues and movements.* Piscataway, NJ: Rutgers University Press.

Cass, N., Shove, E. & Urry, J. (2005). 'Social exclusion, mobility and access'. *The Sociological Review*, 53(3), 539–555.

Clifford, J. (1997). *Routes: Travel and translation in the late twentieth century.* Cambridge, MA: Harvard University Press.

Cook, N. & Butz, D. (2018). 'Gendered mobilities in the making: Moving from a pedestrian to vehicular mobility landscape in Shimshal, Pakistan'. *Social & Cultural Geography*, 19(5), 606–625.

Cook, N. & Butz, D. (2016). 'Mobility justice in the context of disaster'. *Mobilities*, 11 (3), 400–419.

Cook, N. & Butz, D. (2015). 'The dialectical constitution of mobility and immobility: Recovering from the Attabad landslide disaster, Gojal, Gilgit-Baltistan, Pakistan'. *Contemporary South Asia*, 23(4), 388–408.

Cresswell, T. (2010). 'Towards a politics of mobility'. *Environment & Planning D: Society & Space*, 28(1), 17–31.

Cresswell, T. (2006). *On the move: Mobility in the modern Western world.* London: Routledge.

Cresswell, T. (1999). 'Embodiment, power and the politics of mobility: The case of female tramps and hobos'. *Transactions of the Institute of British Geographers*, 24 (2), 175–192.

Elliot, A. & Urry, J. (2010). *Mobile lives.* London: Routledge.

Farrington, J. & Farrington, C. (2005). 'Rural accessibility, social inclusion and social justice: Towards conceptualization'. *Journal of Transport Geography*, 13(1), 1–12.

Featherstone, M., Thrift, N. & Urry, J. (2005). *Automobilities.* Thousand Oaks, CA: Sage.

Fraser, N. (2009). *Scales of justice: Reimagining political space in a globalizing world.* New York: Columbia University Press.

Goggin, G. (2016). 'Disability and mobilities: Evening up social futures'. *Mobilities*, 11 (4), 533–541.

Golub, A. & Martens, K. (2014). 'Using principles of justice to assess the modal equity of regional transportation plans'. *Journal of Transport Geography*, 41, 10–20.

Grieco, M. & Hine, J. (2008). 'Stranded mobilities, human disasters: The interaction of mobility and social exclusion in crisis circumstances'. In S. Bergmann and T. Sager (Eds.), *The ethics of mobilities: Rethinking place, exclusion, freedom and environment* (pp. 65–71). London: Routledge.

Hannam, K., Sheller, M. & Urry, J. (2006). 'Mobilities, immobilities and moorings'. *Mobilities*, 1(1), 1–22.

Ilcan, S. (2013). *Mobilities, knowledge and social justice.* Montreal, QC: McGill-Queen's University Press.

Jackson, J. B. (1984). *Discovering the vernacular landscape.* New Haven, CT: Yale University Press.

Kaufmann, V., Bergman, M. & Joye, D. (2004). 'Motility: Mobility as capital'. *International Journal of Urban & Regional Research*, 28(4), 745–756.

Kenyon, S., Lyons, G. & Rafferty, J. (2002). 'Transport and social exclusion: Investigating the possibility of promoting inclusion through virtual mobility'. *Journal of Transport Geography*, 10(3), 207–219.

Kronlid, D. (2008). 'Mobility as capability'. In T. P. Uteng & T. Cresswell (Eds.), *Gendered mobilities* (pp. 15–33). London: Routledge.

Latour, B. (2005). *Reassembling the social: An introduction to actor-network theory*. New York: Oxford University Press.

Law, J. (1994). *Organizing modernity: Social ordering and social theory*. Oxford: Blackwell.

Lefebvre, H. (1974). *The production of space*. Paris: Anthropos.

Manderscheid, K. (2009). 'Integrating space and mobilities into an analysis of social inequality'. *Distinktion: Scandinavian Journal of Social Theory*, 10(1), 7–27.

Martens, K. (2016). *Transport justice: Designing fair transportation systems*. London: Routledge.

Martens, K. (2012). 'Justice in transport as justice in access: Applying Walzer's "spheres of justice" to the transport sector'. *Transportation*, 39(6), 1035–1053.

Miller, D. (1999). *Principles of social justice*. Cambridge, MA: Harvard University Press.

Montegary, L. & White, M. A. (2015). *Mobile desires: The politics and erotics of mobility justice*. Basingstoke, UK: Palgrave Macmillan.

Mountz, A. (2011). 'Specters at the port of entry: Understanding state mobilities through an ontology of exclusion'. *Mobilities*, 6(3), 317–334.

Mullen, C. & Marsden, G. (2016). 'Mobility justice in low carbon energy transitions'. *Energy Research & Social Science*, 18, 109–117.

Nagel, T. (2005). 'The problem of global justice'. *Philosophy & Public Affairs*, 33(2), 113–147.

Nicholson, J. & Sheller, M. (2016). 'Race and the politics of mobility'. *Transfers*, 6(1), 4–11.

Pereira, R., Schwanen, T. & Banister, D. (2017). 'Distributive justice and equity in transportation'. *Transport Reviews*, 37(2), 170–191.

Rawls, J. (1971). *A theory of justice*. Cambridge, MA: Harvard University Press.

Sabbagh, C. & Schmitt, M. (2016). *Handbook of social justice theory and research*. New York: Springer.

Sager, T. (2006). 'Freedom as mobility: Implications of the distinction between actual and potential travelling'. *Mobilities*, 1(3), 465–488.

Sen, A. (2009). *The idea of justice*. Cambridge, MA: Harvard University Press.

Sen, A. (1993). 'Capability and wellbeing'. In M. Nussbaum and A. Sen (Eds.), *The quality of life* (pp. 30–53). Oxford: Clarendon Press.

Sheller, M. (2015). 'Racialized mobility transitions in Philadelphia: Urban sustainability and the problem of transport inequality'. *City & Society*, 27(1), 70–91.

Sheller, M. (2012). 'The islanding effect: Post-disaster mobility systems and humanitarian logics in Haiti'. *Cultural Geographies*, 20(2), 185–204.

Sheller, M. (2011a). 'Mobility'. *Sociopedia.isa*. doi:10.1177/205684601163.

Sheller, M. (2011b). 'Sustainable mobility and mobility justice: Towards a twin transition'. In M. Grieco & J. Urry (Eds.), *Mobilities: New perspectives on transport and society* (pp. 289–304). London: Routledge.

Sheller, M. & Urry, J. (2006). 'The new mobilities paradigm'. *Environment & Planning A*, 38(2), 207–226.

Skwiot, C. (2009). 'Genealogies and histories in collision: Tourism and colonial contestations in Hawai'i, 1900–1930'. In T. Ballantyne & A. Burton (Eds.), *Moving subjects: Gender, mobility and intimacy in an age of global empire* (pp. 190–210). Chicago, IL: University of Illinois Press.

Talbot, C., Price, S. & Brosnan, S. (2016). 'Inequity responses in nonhuman animals'. In C. Sabbagh & M. Schmitt (Eds.), *Handbook of social theory and research* (pp. 387–406). New York: Springer.

Ureta, S. (2008). 'To move or not to move? Social exclusion, accessibility and daily mobility among the low-income population in Santiago, Chile'. *Mobilities*, 3(1), 269–289.

Urry, J. (2008). 'Moving on the mobility turn'. In W. Canzler, V. Kaufmann & S. Kesserling (Eds.), *Tracing mobilities: Toward a cosmopolitan perspective* (pp. 13–23). London: Routledge.

Urry, J. (2000). *Sociology beyond societies: Mobilities for the twenty-first century.* London: Routledge.

Uteng, T. P. (2006). 'Mobility: Discourses from the non-Western immigrant groups in Norway'. *Mobilities*, 1(3), 431–464.

Vukov, T. (2015). 'Strange moves: Speculations and propositions on mobility justice'. In L. Montegary & M. A. White (Eds.), *Mobile desires: The politics and erotics of mobility justice* (pp. 108–121). Basingstoke, UK: Palgrave McMillan.

Young, I. M. (2002). *Inclusion and democracy.* Oxford: Oxford University Press.

Young, I. M. (1990). *Justice and the politics of difference.* Princeton, NJ: Princeton University Press.

2 Theorizing mobility justice

Mimi Sheller

Introduction

There is growing interest in the concept of 'mobility justice', but no general agreement on what it means. Sometimes the term is assumed to be simply a substitute for 'transportation justice' and mainly concerned with issues of accessibility at the urban scale. This is one important perspective, which builds on an extensive literature within the field of transportation justice that highlights the inequitable distribution of transport access, including racial, ethnic, age, ability and class barriers to mobility. In this chapter, based on my book, *Mobility Justice* (2018), I develop a more holistic theorization of mobility justice that advances debates around transport equity, spatial justice and sustainable cities toward a more comprehensive analysis of intersectional aspects of mobility-related inequities.

The transport justice literature originates in the desegregation efforts of the 1960s civil rights movement in the United States (and their historical precedents going back to the antislavery movement) and egalitarian philosophies of the 1970s that advocated fair use of public space, sometimes expressed as 'the right to the city' (Illich, 1972; Lefebvre, 1974). The right to transportation claimed within the civil rights movement became more explicitly linked to broader issues of equity and fairness in transport planning in the 1990s, especially through the work of Robert Bullard (e.g., Bullard & Johnson, 1997; Bullard, Johnson & Torres, 2000, 2004; Hay, 1993; Hay & Trinder, 1991). Work on transport justice has been consolidated in approaches to transport planning based on principles of justice and measures of accessibility (Martens, 2012, 2016; Pereira, Schwanen & Banister, 2017; Preston & Rajé, 2007). It has also been significantly amplified and connected to broader racial justice issues by recent activist networks such as The Untokening (2016), "a multiracial collective that centers the lived experiences of marginalized communities to address mobility justice and equity," and the newly formed activist organization People for Mobility Justice (peopleformobilityjustice.org).

However, recent work on mobility-related questions of justice incorporates a wider range of issues beyond transportation or urban accessibility (e.g., Cook & Butz, 2016; Sheller, 2015a, 2015b; Vukov, 2015). It may include, for example, uneven impacts of transport-related pollution on public health linked more broadly to 'environmental justice', other harmful social and environmental impacts of automobility linked to 'climate justice', exclusions from public space and the 'right to the city' linked to 'spatial justice', conflicts over border regimes and the rights of migrants linked to 'migrant justice' and uneven effects of climate change and disaster-related (im)mobilities linked to 'climate justice'. While much attention is given to the problem of developing sustainable and accessible urban transportation, I argue that a comprehensive theory of mobility justice should address multiple scales of mobility and multiple approaches to justice in all of these areas.

Unlike existing approaches to sustainable mobility transitions that focus on transportation alone, my approach highlights the power relations among bodies, transport systems, urban spaces, national and transnational mobility regimes and systems of planetary urbanization and their associated logistics infrastructures (Brenner & Schmid, 2015). Questions of mobility-related injustices, therefore, relate beyond transportation and accessibility to: (a) embodied relations of race, gender, age, disability, sexuality, etc. that inform uneven freedom of movement; (b) uneven (non) urban spaces and unequal transport infrastructure and accessibility at the local and regional level (e.g., unequal distributions of automobility, public transit, biking access, ride-sharing, etc.); (c) national mobility regimes related to the differential mobilities of borders, migration, human trafficking, asylum seeking, detention, etc.; and (d) infrastructural issues relating to the uneven circulation of goods, resources and energy in a planetary urbanization system.

This holistic approach to mobility justice places debates over sustainable transportation, low-carbon transitions and what are usually thought of as localized urban transportation justice issues in the wider context of many different unequal mobility regimes extending from face-to-face bodily relations to extensive planetary circulation. It focuses attention on issues of sustainable and just transportation in cities, but also micro-level embodied spatial relations of (im)mobility, (dis)ability and the right to the city as well as macro-level transnational relations of travel, migration, borders and uneven circulation of goods and resources. Mobilities research offers a transversal means of conjoining urban studies, transport planning, tourism studies, migration studies and other fields. By focusing on cross-cutting problems of the politics of uneven (im)mobilities through a common prism, we can begin to show how they not only intersect, but refract and intensify each other in multiple ways simultaneously.

A more comprehensive theory of mobility justice can also address the combined crises of climate change, unsustainable urbanization and global migrations as part of a common phenomenon of uneven mobilities that impact everyday life. Mobility justice is an overarching concept for thinking about how power and inequality inform the governance and control of movement, shaping the patterns of unequal mobility and immobility in the circulation of people, resources and information.

To achieve this multifaceted approach, we first need to rethink the theorization of justice within a kinopolitical mobile ontology (Nail, 2015, 2016). A focus on the cross-cutting scales of (im)mobilities is crucial, but so is a cross-disciplinary theorization of mobilities based on a mobile ontology. Some research has begun to theorize the politics of mobility as "kinopolitical", a concept especially developed by Thomas Nail. Critical mobilities research highlights a range of power relations of (im)mobility (Bærenholdt, 2013; Jensen, 2011), spatial barriers that create forms of social exclusion (Blomley, 2011; Cass, Shove & Urry, 2005) and the problems of differential mobilities and the 'mobility poor' (Cresswell, 2010). But it has yet to be elaborated how the making of (im)mobile subjects within a kinopolitical framework moves beyond existing notions of justice, including transport justice.

In the following three sections I offer an overview of mobilizing justice, moving beyond transport justice towards mobility justice. The new mobilities paradigm develops a mobile ontology as a way of apprehending complex realities that shape space, time and movement through differential processes of mobilization, immobilization, speeding up and slowing down, holding still and releasing. A mobile ontology, in which *movement is primary as a foundational condition of being, space, subjects and power*, helps us rethink the philosophy of justice.

Mobilizing justice

The new mobilities paradigm crosses micro, meso and macro levels, ranging from inter-human bodily relations to transportation and street design, to urban problems and extended infrastructural space, to fossil fuels and planetary mobilities. It asks how relations of (im)mobility, speed and power are produced within and through social and political practices. It also attends to concepts like differential mobility, uneven mobilities, 'motility' or potential mobility and mobility capabilities, and to questions of power, justice and mobility rights (e.g., Bærenholdt, 2013; Faulconbridge & Hui, 2016; Flamm & Kaufmann, 2006; Kellerman, 2012; Kronlid, 2008; Sheller, 2015b).

Control over mobility is a form of power with deep historical roots. As Kotef (2015, pp. 10–11) elaborates, "[l]iberal democracies have always operated in tandem with regimes of deportation, expulsion and expropriation, as well as confinement and enclosure, implementing different rationalities of rule to which colonized, poor, gendered and racialized subjects were subjected". Liberal models of subjectivity and freedom depend on the self-regulation of movement, such that "the liberal subject is essentially a moving subject, and her first and most fundamental freedom is freedom of movement" (Kotef, 2015, p. 58). Kotef traces the history of liberalism as a regime of movement in which the "free movement of some [people] limits, hides, even denies the existence of others" (2015, p. 54). As this liberal subject and citizen forms through rights to (well-regulated) movement, protected by the territorial state, it simultaneously produces an 'other' who suffers exclusion, enclosure, incarceration and violence (Kotef, 2015, pp. 54, 37, 58).

Differential mobilities, therefore, are fundamental to forms of power that make classed, racial, sexual, able-bodied, gendered, citizen and noncitizen subjects, as well as to forms of resistance and countermoves (Cresswell, 2016; McKittrick, 2006). Wider histories of liberal citizenship suggest that the right to unhindered mobility helps to produce the nation-state, even while some citizens' mobility is "constantly hindered", as Cresswell (2006, p. 161) notes in the US context: "Arab Americans stopped at airport immigration, Hispanic Americans in the fields of American agri-business or African Americans 'driving while black'". Thus, liberal notions of mobility as freedom, and of the free individual who is self-moving and autonomous, need to be revisited. In this view, "space becomes political via the movements it allows and prevents, and the relations that are formed or prevented via these im/mobilities" (Kotef, 2015, p. 114).

The struggle for mobility justice is a core political gradient or fault line encompassing social and political struggles over space, access, movement and the power relations that mobilities enable or disrupt. Like approaches to climate justice that combine distributive, procedural, participatory and capability dimensions of justice, a multipronged approach is necessary "because in order for people to gain access to material (distributional) social goods, they must be valued and included (recognition) through access to society's decision-making institutions (participatory parity) and society's basic institutions (capabilities)" (Harlan, Pellow & Roberts, 2015, p. 136), as historical damages are redressed through restorative justice. Because unequal mobilities are the mechanism for denying capabilities for access, participation and inclusion, mobility justice is fundamental to achieving all of the above. Most importantly, these mobility inequities extend from bodily restrictions and disabling environments to transport exclusions and uneven infrastructure to unequal access to resources and energy, and unequal exposures to harms and pollution.

Thus, I want to mobilize theories of justice so as to encompass a wider range of concerns within the overarching concept of 'mobilities' (this is what community-based organizers in The Untokening achieve through their "Principles of Mobility Justice", 2016). We must consider how to combine political movements for accessibility and bodily freedom of movement with movements for equitable infrastructures and spatial designs that support a shared mobility commons, movements for fair and just forms of sustainable transport and ecological urbanism that reduce environmental harms and movements for the equitable global distribution of natural benefits and harms, including human and nonhuman rights to move and to dwell. Mobility (in) justices cross these fields of action, flowing from one into another, enfolding them or ricocheting across them. Indeed, mobility injustices do not occur after entities 'enter' a space (i.e., after travelers get into a vehicle, people gather on a city street or migrants enter a new country), but *are the process through which unequal spatial conditions and differential subjects are made.*

As Kotef (2015, p. 114) argues, "movement is one of the attributes of political spaces: political spaces are often moving spaces. Movement thereby becomes primary within the anatomy of political spheres". This is true within the internal politics of a city or state as well as in external political struggles over borders, citizenship, voice and exit, incarceration and freedom. Indeed, mobility is the basis upon which the distinction between public and private spheres rests and, hence, is the condition for public assembly and the existence of politics (Emirbayer & Sheller, 1999; Sheller & Urry, 2003). This is why mobility justice requires a mobile ontology and demands a kinopolitical approach that challenges liberal conceptualizations of freedom of movement and exceeds distributive notions of justice based simply on increasing accessible transportation.

Beyond transport justice

An account of the distinctiveness of 'mobility justice' begins by examining the limitations of existing theoretical approaches to justice within the transport justice literature. Acknowledging that many such arguments lack solid philosophical underpinnings, Pereira, Schwanen and Banister (2017) explore the applicability of various theories of justice for thinking about transport. They consider the limitations of utilitarian, liberal and intuitionist theories of justice and then develop an approach combining Rawlsian egalitarianism and capability approaches. Emerging applications of capabilities approaches not only emphasize the need for fairness in the distribution of transport and accessibility, but also call for greater attention to justice in transportation decision-making and participatory processes (Hananel & Berechman, 2016; Kronlid, 2008).

Pereira, Schwanen and Banister (2017, p. 171) draw on theorists such as Nancy Fraser (1995), William Kymlicka (2002) and Iris Marion Young (1990) to define justice

> as a broad moral and political ideal that relates to (a) how benefits and burdens are distributed in society (distributive justice); (b) the fairness of processes and procedures of decision and distribution (procedural justice); and (c) the rights and entitlements which should be recognized and enforced.

They note that Rawls' (1999, 2001) approach to justice rests on two fundamental principles. First, and having priority, is the principle that "the rules defining individuals' rights and liberties ought to apply equally to everyone and that individuals should have as much freedom as possible as long as this does not infringe the freedom of others" (Pereira, Schwanen & Banister, 2017, p. 174). Second, the distribution of social goods (e.g., wealth, opportunities, the bases of self-respect) should "simultaneously (a) derive from a situation of fair equality of opportunity, and (b) work to the benefit of the least advantaged members of society" (Pereira, Schwanen & Banister, 2017, p. 174).

According to this 'difference principle', inequalities of opportunity and arbitrary effects may exist; however, policies should rest on a distributive rule that seeks to maximize the minimum level of primary goods for the least well-off. In relation to transportation, *distributive justice* implies the idea of equitable distribution of and access to mobility, but also the equitable distribution of the risks and benefits associated with mobility infrastructures. However, given the resource and spatial constraints on mobility and the conflict between different exercises of mobility, mobility justice cannot simply involve an extension of maximum mobility to all; more people having access to cars, airplanes and highways would increase congestion, greenhouse gases and other environmental pollution harms.

Critics of distributive justice also note that some forms of participation in movement constitute barriers to other activities or choices. For example, too many people exercising the freedom of automobility prevents others from walking or biking on roadways. Moreover, distributive justice is insufficient to address mobility injustices because it holds space still and treats accessibility simply as the movement from A to B for other purposes. It thus fails to engage with how the exercise of mobility is always an exercise of power within a kinopolitical system. Automobility, for example, is not merely a problematic form of transportation that can be replaced by more accessible public and active transport; it also concerns the spatial orders around speed, its cultural valuation and uneven distribution and its relation to climate change, fossil fuel infrastructures, suburbanization and the difficulty of global post-carbon transitions (Dennis & Urry, 2009; Sheller, 2014). There are complex relations between consumption and climate change that involve status, class and culture, not to mention embedded institutions, routines, practices, time use and lifestyles, which complicate the problem of low-carbon technological transitions (Ehrhardt-Martinez & Schor, 2015).

Pereira, Schwanen and Banister (2017) also engage critiques of Rawls by capabilities approach (CA) advocates, such as Amartya Sen (2009) and Martha Nussbaum (2011). For Sen, the distribution of resources or primary goods should not be an end in itself, because it is incapable of recognizing the diversity of human needs and preferences, which concern more deeply valued ends and aspirations. People's capacities are shaped by their opportunities, which are in turn shaped by both their internal capabilities and their external environment, including societal structures that might constrain a person's 'functionings'. Thus, CA implies the guarantee of certain basic capabilities and minimum thresholds. Freedom of movement would be one such capability. The problem arises, however, if we need to put a maximum threshold on some people's freedoms, once their basic needs have been met, in order to limit harm to others' capabilities. Should there be a limit on freedom of movement?

According to Nussbaum, a minimally just society requires entitlements to life, bodily health and integrity, freedom of movement and political and material control over one's environment, amongst other capabilities. Mobility, in this perspective, is the freedom to move in order to make desired goods, services or ends accessible. But once accessibility is met, should freedom of movement in and of itself be valued? Too much movement may be socially harmful insofar as it creates pollution, congestion and sprawl. This is why transport justice scholarship (e.g., Martens, 2016) focuses instead on *accessibility*, "conceptualized as the ease with which persons can reach places and opportunities from a given location, and be understood as the outcome of the interplay of characteristics of individuals, the transport system, and land use" (Pereira, Schwanen & Banister, 2017, p. 177).

This notion of accessibility leads Martens (2016) to develop three rules of "people-centered transport planning": (a) start from the people, not the system; (b) assess transport interventions on their effectiveness in bringing people above the threshold, rather than their efficiency; and (c) finance transport through income-based accessibility insurance schemes, not car-based forms of taxation. Focusing on accessibility helps shift attention away from increasing mobility for all, which leads to perpetual growth in movement and the associated problems of pollution and congestion. However, it still holds in place the beginning and endpoints of such access journeys. In other words, it treats space as a container for mobile activities, and inequalities as preexisting conditions not subject to change. Policies promoting accessibility also leave unanswered basic philosophical questions about how mobility is connected to ideas of freedom, individualism and liberalism that have historically shaped and structured uneven spatial relations (and racial/gender/sexual relations).

Pereira, Schwanen and Banister (2017) argue that applying the Rawlsian difference principle to transport implies prioritizing public transit, walking and cycling in dense urban areas and subsidizing car ownership in low-density, nonurban areas. They argue for expanding the idea of mobility as a capability "into an understanding of accessibility as a combined capability" such that people are able to convert transport resources into access to "activities that are essential for meeting basic needs, such as food stores, education, health services and employment opportunities" (Pereira, Schwanen & Banister, 2017, p. 182). Like Martens, they argue that their egalitarian capabilities approach depends on defining minimum thresholds of accessibility and establishing policies through legitimate democratic processes to guarantee those thresholds. Even so, they note the difficulties involved in measuring accessibility, the tensions between focusing on person-based and place-based measures and the challenges to the underlying ideas of agency and freedom of choice given the possible need for limitations on some forms of mobility to ensure greater overall fairness.

Yet, Pereira, Schwanen and Banister (2017) set aside feminist theories of justice (e.g., Young, 1990), which pay more attention to questions of recognition, embodiment and the exclusion of some people from deliberative processes, and movements concerned with racial justice or critical disabilities. How can we achieve democratically legitimate planning around mobility and accessibility if existing patterns of uneven mobilities empower some groups while disempowering others? It is precisely at the intersection of bodily differences in capabilities and entrenched spatial injustices that we need to develop a more intersectional and multi-scalar theory of mobility justice – one that moves beyond transport justice. We need to focus on the intersection of multiple scales and temporalities of mobile relations and processes and to develop more collective, non-individualistic and commons-based understandings of mobility. In the following section I begin to outline such an approach.

Towards mobility justice

The mobilities of the most privileged groups, including the use of large vehicles, extensive air travel and greater energy consumption, contribute to global warming that has greater impacts on lower-income regions of cities and the world. How can we connect the theorization of mobility justice to the multiple scales at which liberal freedoms of mobility are actually harmful? First, at the scale of the body, we know that poor and vulnerable populations experience the greatest harm, injury and death from unjust mobility systems. The poor are most likely to be excluded from access to convenient and safe forms of transportation, and they suffer the highest rate of pedestrian deaths from motor vehicle collisions. They are exposed to greater air pollution and health impacts of climate change. Gender, age, race and disability also restrict movement in many ways. The least 'able-bodied' face major hurdles in accessing urban space and moving around, being 'disabled' by built environments that prevent assisted mobility.

Second, urban mobility regimes articulate with transnational regimes that bring immigrant populations into urban peripheries and suburbs, which typically lack good transit connections. Noncitizens often have to take buses with inadequate stops and schedules and risk dangerous walks and bike rides along unsafe roadways with insufficient cycling infrastructure in order to get to jobs that lack employer subsidies for transportation, in contrast to those who arrive by car along subsidized highways to free parking places and tax breaks. Mobility and race have intersected historically, and they intersect today in unequal relations of power that make mobility racially loaded in particular moments and places, while also making racial processes and racialized spaces and identities (including whiteness) deeply contingent on differential mobilities (Cresswell, 2016).

Third, there are crucial ways in which uneven mobilities involve infrastructure spaces and 'dispositions' (Easterling, 2015) that utilize and bring into being extended operational landscapes of urbanization at a planetary scale, such as mining, fracking, pipelines and hydropower (Arboleda, 2016). Technologies of mobility embody (and lock in) energy in their production, moment of use (doing particular kinds of work) and relation to specific infrastructural 'moorings' predicated on unequal access to particular forms of energy such as liquid petroleum or the energy embedded in the car itself (Hannam, Sheller & Urry, 2006). Urban systems and transnational mobilities produce technological regimes that embed particular energy cultures as assemblages of energy generation, metals, vehicles and objects, which then become materialized in everyday life.

American suburbanization, for example, is deeply dependent on access to oil. Matthew Huber (2013, p. xi) argues that "[o]il is a powerful force not only because of the material geographies of mobility it makes possible, but also because its combustion often accompanies deeply felt visions of freedom and individualism". Oil extraction produced a way of life rooted in automobility, petrochemicals, plastics and industrial agriculture. Suburbanization was a structuration of space, but it also produced what Raymond Williams calls a "structure of feeling" based on fossilized "mobile privatization" (quoted in Huber, 2013, pp. 23, 74). One implication of suburban American lifestyles, histories of racialized urban segregation and the dependence of both on cheap fossil fuels is that before we can address the injustices of hydrocarbons, we first need to stop disregarding our own involvement with them. White middle-class Americans must acknowledge their own responsibility as 'high emitters' of carbon dioxide. Suburban dwellers must acknowledge their role in the splintered provision of unequal mobilities, the associated deficits in urban accessibility and the exposure of the mobility poor to greater climate risk and vulnerability. Mobility injustice, in other words, relates to how we dwell as much as how we move.

The theorization of network capital is useful for envisioning the self-determination of mobility *and* immobility as forms of power: a capability to move or be still and to make others move. Elliott and Urry (2010, pp. 10–11) define network capital as a combination of capacities to be mobile, including appropriate documents, money and qualifications; access to networks at-a-distance; physical capacities for movement; location-free information and contact points; access to communication devices and secure meeting places; access to vehicles and infrastructures; and time and other resources for coordination. There is an uneven distribution of these capacities for potential movement in relation to the surrounding physical, social and political affordances for movement, but such uneven network capital also distributes harms unevenly.

Stephen Gössling and Scott Cohen (2014) propose that there are 'taboos' on particular transport policies, making them impossible to even mention because they are so politically risky. Especially significant is the political unwillingness to acknowledge that

> there are huge differences in the power geometries of individual mobility (e.g., Gössling et al., 2009; Schäfer et al., 2009), with a minor share of highly mobile travelers being responsible for a significant share of the overall distances travelled, as well as emissions associated with this transport.
>
> (Gössling & Cohen, 2014, p. 200)

A small proportion of privileged travelers (generally white men earning high incomes) engage in frequent and long-distance travel – even 'binge flying' – and drive large-engine luxury cars, producing hugely disproportionate amounts of greenhouse gases and other pollutants (and taking up more urban space for these activities).

Mobility justice, therefore, requires distributive justice achieved through increasing accessibility within existing urban spatial forms, but also a deeper deliberation over substantive values to determine which activities should be protected (e.g., funding for and access to public transport systems) and which should be reduced (e.g., free parking, subsidies for automobiles, frequent air travel). If 'motility' is defined as "the manner in which an individual or group appropriates the field of possibilities relative to movement and uses them" (Kaufmann & Montulet, 2008, p. 45), then mobility justice involves deliberating over where to draw the limits of the field of possibilities and of particular forms of appropriation. Deliberation over mobility justice, therefore, is not simply about extending the means of transport or even expanding accessibility, but also requires deciding on wider distributions of transportation harms.

Such 'deliberative justice' should take into account the reasonable possibility that public input could influence the outcomes of a decision. Environmental justice theorists (e.g., Schlosberg, 2007; Shrader-Frechette, 2005), for example, argue that efforts must be made to address preexisting power inequalities among participants using deliberative processes rooted in egalitarian principles (rather than pluralistic processes in which private interests compete). Deliberation first requires 'recognition'. Decision makers need to acknowledge the legitimacy of community members' participation and respect their input as an important and relevant contribution to decision-making. Recognition would also require lifting the taboos on certain discussion topics.

Deliberation and recognition extend into matters of 'procedural justice', defined as meaningful participation of affected populations in the governance of mobility systems, which needs to extend beyond the local or even urban scale. Fair procedures for democratic deliberation require at least the following: (a) access to information; (b) substantial understanding (requiring community-based participatory production of knowledge); and (c) informed consent based on 'local knowledge' (rather than just expert knowledge), which must be grounded in measures to support capabilities for participation and understanding.

Climate justice frameworks also call for such broad-based stakeholder participation and a transformative approach to socio-ecological relations writ large (Dryzek, Norgaard & Schlosberg, 2013). Schlosberg (2012, p. 446), building on CA theories of justice (Nussbaum, 2011; Sen, 2009) and political theories of recognition (Fraser, 1997; Young, 1990), argues that a capabilities approach to collective normative frameworks can "bring social and political recognition of specific and local vulnerabilities and the effects of climate change on the basic needs of human beings in various places and under different conditions".

Procedural justice ensures the participation of disempowered groups in deliberation and decision-making processes. Such 'participatory design' is crucial to realizing more just socio-technological systems. As entire urban populations become vulnerable to warfare, natural disasters and climate change that threaten access to water, energy and food, we also need to conceive of mobility rights in terms of 'restorative justice'. Insofar as 'kinetic elites' have contributed the most to greenhouse gas emissions (Ehr-hardt-Martinez & Schor, 2015), they should also take responsibility for those harms on groups that have contributed little to global warming. Therefore, we need new approaches to admitting responsibility, realizing truth and reconciliation and making reparations to those harmed by climate change (or urban sprawl, excessive resource use and pollution) due to the excessive mobility of others.

Gwen Ottinger (2013) cautions that relevant information related to human environmental impacts simply may not yet exist, as scientific knowledge production is emergent and changing. If entire knowledge systems have yet to come into existence, then including local knowledge into deliberations over equitable and sustainable mobilities is an appropriate move that goes beyond procedural inclusion and participation parity. Therefore, "procedural justice should include proactive knowledge production to fill in knowledge gaps and ongoing opportunities for communities to consent to the presence of hazards as local knowledge emerges and scientific knowledge changes" (Ottinger, 2013, p. 250). Addressing knowledge gaps and including alternative modes of knowing (such as Indigenous ontologies) lead to 'epistemic justice', which involves recognizing and creating new facts, forms of knowledge and ways of reconciling seemingly incommensurable ways of knowing.

These justice-related processes lead us beyond a narrow focus on transportation and accessibility because principles of recognition and procedural justice allow *other relevant topics* to be raised, such as the environmental harms associated with oil drilling, oil pipelines and hydraulic fracturing that support automobility. They might also raise questions about the impacts of hydroelectric dams, land appropriation and the displacement of Indigenous populations, and about the use of fossil fuels in agro-industrial food production and circulation and their potential social and ecological harms. In this framing, not only is accessibility to transportation, jobs, education and political participation a crucial mobility justice issue, but so too are conflicts over 'fast food' and forest clearance to raise cattle for hamburgers; the Dakota Access Pipeline, Canadian tar sands exploitation and drilling for oil in the Arctic; and rights to refuge, asylum and migration driven by environmental harm.

These wider kinopolitical issues do not emerge as concerns within circumscribed discussions of transport justice, but they are crucial 'mobility justice' concerns. This scalar shift is fundamental to broadening the political framing and procedural issues that inform urban transport policy and planning, including *who* is recognized as a participant, *what* is recognized as a legitimate topic of deliberation and *where* (and at what scale) conflicts should be resolved.

Conclusion

This conceptualization of mobility justice draws on a mobile ontology that incorporates entangled scales and kinopolitical concepts. Approaching justice as a more mobile concept shifts the discussion of mobility justice beyond transport justice approaches that focus on day-to-day policymaking around accessibility and infrastructure planning (important though that is) and beyond ideas of spatial justice that focus on the city as a specific location and delimited urban forms. Such nationally based, urban-focused mobility politics have been too easily co-opted into projects of urban boosterism and gentrification through spatial fixes of the entrepreneurial city. Improved transport infrastructure simply displaces the politically marginalized and exacerbates mobility poverty.

Crucially, this broader perspective on mobility justice requires more sustained attention both to micro-level embodied differential (im)mobilities, that are always racialized and gendered relations to space, and to macro-level colonial histories and postcolonial understandings of the historical antecedents of contemporary forms and patterns of global (im)mobilities. The concept of mobility justice, therefore, draws on insights from arenas such as transport justice, racial justice and environmental justice, but also differs from them in specifying a mobile ontology that pertains to all forms of movement and by focusing on wide-ranging techniques for managing different kinds of (im)mobilities and infrastructures.

Deliberation over mobility justice, therefore, pertains not simply to expanding transport infrastructures or even accessibility, but also to the cultural meanings and hierarchies surrounding various means of mobility, infrastructures for mobility, their valuation and who determines value, relevant facts and meaning. Even more radically, it would take into account how kinopolitics shape space, subjects and difference in the first place.

I hope that a more robust, multidimensional and historically embedded theory of mobility justice – drawing on colonial, corporeal and planetary histories and interrelations – can help us combine political efforts and social movements that have heretofore been separated into a powerful unity. The promise of this conceptualization of mobility justice is that it can bring together embodied movements for social justice (combining class, race, gender, disability and sexuality), struggles for transport justice and accessibility, arguments for the right to the city and spatial justice, movements for migrant and Indigenous rights, de-colonial movements, climate justice and global equity – all under one common framework. This intersectional, multi-scalar approach can help build more politically effective alliances and gain greater leverage over the urgent crises of the Anthropocene, which will shape our ways of moving, mobilizing, dwelling and living in the near and distant future.

References

Arboleda, M. (2016). 'Spaces of extraction, metropolitan explosions: Planetary urbanization and the commodity boom in Latin America'. *International Journal of Urban & Regional Research*, 40(1), 96–112.

Bærenholdt, J. (2013). 'Governmobility: The powers of mobility'. *Mobilities*, 8(1), 20–34.

Blomley, N. (2011). *Rights of passage: Sidewalks and the regulation of public flow*. New York: Routledge.

Brenner, N. & Schmid, C. (2015). 'Towards a new epistemology of the urban?' *City*, 19 (2–3), 151–182.

Bullard, R. & Johnson, G. (1997). *Just transportation: Dismantling race and class barriers to mobility*. Gabriola Island, BC: New Society Publishers.

Bullard, R., Johnson, G. & Torres, A. (2004). *Highway robbery: Transportation racism and new routes to equity*. Cambridge, MA: South End Press.

Bullard, R., Johnson, G. & Torres, A. (2000). 'Dismantling transportation apartheid: The quest for equity'. In R. Bullard, G. Johnson & A. Torres (Eds.), *Sprawl city* (pp. 39–68). Washington, DC: Island Press.

Cass, N., Shove, E. & Urry, J. (2005). 'Social exclusion, mobility and access'. *The Sociological Review*, 53(3), 539–555.

Cook, N. & Butz, D. (2016). 'Mobility justice in the context of disaster'. *Mobilities*, 11 (3), 400–419.

Cresswell, T. (2016). 'Black moves: Movements in the history of African-American mobilities'. *Transfers*, 6(1), 12–25.

Cresswell, T. (2010). 'Towards a politics of mobility'. *Environment & Planning D: Society & Space*, 28(1), 17–31.

Cresswell, T. (2006). *On the move: Mobility in the modern Western world*. London: Routledge.

Dennis, K. & Urry, J. (2009). *After the car*. London: Routledge.

Dryzek, J., Norgaard, R. & Schlosberg, D. (2013). *Climate-challenged society*. Oxford: Oxford University Press.

Easterling, K. (2015). *Extrastatecraft: The power of infrastructure space*. New York: Verso.

Ehrhardt-Martinez, K. & Schor, J. (2015). 'Consumption and climate change'. In R. Dunlap & R. Brulle (Eds.), *Climate change and society: Sociological perspectives* (pp. 93–126). New York: Oxford University Press.

Elliott, A. & Urry, J. (2010). *Mobile lives*. London: Routledge.

Emirbayer, M. & Sheller, M. (1999). 'Publics in history'. *Theory & Society*, 28(1), 145–197.

Faulconbridge, J. & Hui, A. (2016). 'Traces of a mobile field: Ten years of mobilities research'. *Mobilities*, 11(1), 1–14.

Flamm, M. & Kaufmann, V. (2006). 'Operationalizing the concept of motility: A qualitative study'. *Mobilities*, 1(2), 167–189.

Fraser, N. (1997). *Justice interruptus: Critical reflections on the 'postsocialist' condition*. New York: Routledge.

Fraser, N. (1995). 'Recognition or redistribution? A critical reading of Iris Marion Young's *Justice and the Politics of Difference*'. *Journal of Political Philosophy*, 3(2), 166–180.

Gössling, S., Ceron, J., Dubois, G. & Hall, C. (2009). 'Hypermobile travellers'. In S. Gössling & P. Upham (Eds.), *Climate change and aviation* (pp. 131–149). London: Earthscan.

Gössling, S. & Cohen, S. (2014). 'Why sustainable transport policies will fail: EU climate policy in the light of transport taboos'. *Journal of Transport Geography*, 39, 197–207.

Hananel, R. & Berechman, J. (2016). 'Justice and transportation decision-making: The capabilities approach'. *Transport Policy*, 49, 78–85.

Hannam, K., Sheller, M. & Urry, J. (2006). 'Mobilities, immobilities and moorings'. *Mobilities*, 1(1), 1–22.

Harlan, S., Pellow, D. & Roberts, J. (2015). 'Climate justice and inequality'. In R. Dunlap & R. Brulle (Eds.), *Climate change and society: Sociological perspectives* (pp. 127–163). New York: Oxford University Press.

Hay, A. (1993). 'Equity and welfare in the geography of public transport provision'. *Journal of Transport Geography*, 1(2), 95–101.

Hay, A. & Trinder, E. (1991). 'Concepts of equity, fairness and justice expressed by local transport policymakers'. *Environment & Planning C: Government & Policy*, 9 (4), 453–465.

Huber, M. (2013). *Lifeblood: Oil, freedom and the forces of capital*. Minneapolis, MN: University of Minnesota Press.

Illich, I. (1972). *Energy and equity*. Retrieved from: www.preservenet.com/theory/Illich/EnergyEquity/Energy%20and%20Equity.htm. Accessed 30 November 2017.

Jensen, A. (2011). 'Mobility, space and power: On the multiplicities of seeing mobility'. *Mobilities*, 6(2), 255–271.

Kaufmann, V. & Montulet, B. (2008). 'Between social and spatial mobilities: The issue of social fluidity'. In W. Canzler, V. Kaufmann & S. Kesselring (Eds.), *Tracing mobilities: Towards a cosmopolitan perspective* (pp. 37–56). Farnham: Ashgate.

Kellerman, A. (2012). 'Potential mobilities'. *Mobilities*, 7(1), 171–183.

Kotef, H. (2015). *Movement and the ordering of freedom: On liberal governances of mobility*. Durham, NC: Duke University Press.

Kronlid, D. (2008). 'Mobility as capability'. In T. P. Uteng & T. Cresswell (Eds.), *Gendered mobilities* (pp. 15–34). Aldershot: Ashgate.

Kymlicka, W. (2002). *Contemporary political philosophy: An introduction*, 2nd ed. Oxford: Oxford University Press.

Lefebvre, H. (1974). *The production of space*. Cambridge: Blackwell.

Martens, K. (2016). *Transport justice: Designing fair transportation systems*. London: Routledge.

Martens, K. (2012). 'Justice in transport as justice in accessibility: Applying Walzer's "spheres of justice" to the transport sector'. *Transportation*, 39(6), 1035–1053.

McKittrick, K. (2006). *Demonic grounds: Black women and the cartographies of struggle*. Minneapolis, MN: University of Minnesota Press.

Nail, T. (2016). *Theory of the border*. Oxford: Oxford University Press.

Nail, T. (2015). *The figure of the migrant*. Stanford, CA: Stanford University Press.

Nussbaum, M. (2011). *Creating capabilities: The human development approach*. Cambridge, MA: Harvard University Press.

Ottinger, G. (2013). 'Changing knowledge, local knowledge and knowledge gaps: STS insights into procedural justice'. *Science, Technology & Human Values*, 38(2), 250–270.

Pereira, R. H. M., Schwanen, T. & Banister, D. (2017). 'Distributive justice and equity in transportation'. *Transport Reviews*, 37(2), 170–191.

Preston, J. & Rajé, F. (2007). 'Accessibility, mobility and transport-related social exclusion'. *Journal of Transport Geography*, 15(3), 151–160.

Rawls, J. (2001). *Justice as fairness: A restatement*. Cambridge, MA: Harvard University Press.

Rawls, J. (1999). *A theory of justice*. Cambridge, MA: Harvard University Press.

Schäfer, A., Heywood, J. B., Jacoby, H. D. & Waitz, I. A. (2009). *Transportation in a climate-constrained world*. Cambridge, MA: MIT Press.

Schlosberg, D. (2012). 'Climate justice and capabilities: A framework for adaptation policy'. *Ethics & International Affairs*, 26(4), 445–461.

Schlosberg, D. (2007). *Defining environmental justice: Theories, movements and nature*. Oxford: Oxford University Press.

Sen, A. (2009). *The idea of justice*. Cambridge, MA: Harvard University Press.

Sheller, M. (2018). *Mobility justice: The politics of movement in an age of extremes*. London: Verso.

Sheller, M. (2015a). 'Racialized mobility transitions in Philadelphia: Urban sustainability and the problem of transport inequality'. *City & Society*, 27(1), 70–91.

Sheller, M. (2015b). 'Uneven mobility futures: A Foucauldian approach'. *Mobilities*, 11(1), 15–31.

Sheller, M. (2014). *Aluminum dreams: The making of light modernity*. Cambridge, MA: MIT Press.

Sheller, M. & Urry, J. (2003). 'Mobile transformations of "public" and "private" life'. *Theory, Culture & Society*, 20(3), 107–125.

Shrader-Frechette, K. (2005). *Environmental justice: Creating equality, reclaiming democracy*. Oxford: Oxford University Press.

The Untokening. (2016). 'Untokening 1.0: Principles of mobility justice'. Retrieved from http://untokening.com.

Vukov, T. (2015). 'Strange moves: Speculations and propositions on mobility justice'. In L. Montegary & M. White (Eds.), *Mobile desires: The politics and erotics of mobility justice* (pp. 108–121). New York: Palgrave Macmillan.

Young, I. M. (1990). *Justice and the politics of difference*. London: Routledge.

Part II
Developing mobility justice

Justice and mobility governance

3 Aeromobility justice
A global institutional perspective

Weiqiang Lin

Introduction

Air travel perhaps represents one of the most salient cases where spatial mobility plays an influential role in differentiating the socioeconomic lives of people and structuring society's hierarchies. By unevenly sorting those who move and those who do not, as well as dictating how easily they move, it invokes serious questions of justice or, more specifically, *mobility* justice. As Sheller (2013, p. 186) writes, an approach highlighting mobility justice "focuses our attention on who is able to exercise rights to mobility and who is not capable of mobility within particular situations". In aeromobility's case, the discrepancy between those who move and those who do not is not only common, but resilient, not least due to the specialty of aviation as a technical mode of transport that not everyone – or, indeed, every place – can get access to. Amplifying the effects of this differential access are immense economic potentials and social possibilities associated with flying that might be gained, or lost, by virtue of one's being unevenly plugged into the world's aeromobile regimes (Cwerner, 2009). Ingrained within aeromobility, therefore, is a persistent politics that implicitly determines, or rather splinters, the futures of people in profound ways, acting as sites of inequality that both open up and curtail individuals' life chances at the planetary scale. In an age when being globally connected is a hallmark of success, the undemocratic spread of aeromobility is an even more pressing issue to be addressed.

A critical stance that interrogates the implications of flying on people's rights to spatial access is not new. Since the mid-2000s, when the 'new' mobilities paradigm came to prominence, scholars have sought to inject aeromobility with a sense of sociocultural relevance (Adey, Budd & Hubbard, 2007), effectively taking the study of air transport away from economistic analyses of hubbing patterns, traffic volumes and infrastructural building. While such research has yielded important insights on the inequities plaguing different groups of travellers (e.g., Adey, 2010; Martin, 2011), the institutional mechanisms by which flying is systematically and globally set up as a geographically variegated affair among *nations* have largely remained unexplored. What proves even more of a black box are the governance apparatuses, laws and logics that give rise to uneven situations of mobility within the international context, even before one can speak of the finer distinctions of mobility among individuals. It follows, then, that there is a need to delve deeper into the governance practices and regulatory frameworks in (aero)mobilities production that precipitate these imbalanced ways of moving at the global scale (Lin, 2018). In pursuit of a less conventional kind of mobility justice, such a focus helps attribute part of the problem of discrepant mobilities to international institutional actors and their actions. It avers that the source of Sheller's (2013) "islanding effect" – or the marooning of certain groups of people from mobility – is not an acephalous one, but the outcome of complex choreographed efforts among government-level decision makers to exclude, marginalize and slow down at a macrostructural level.

This chapter dissects the institutional practices latent within global aviation as a test case to demonstrate the salience of global governance processes in considerations of mobility justice. Admittedly, the aviation industry offers a highly distinctive model for examination, involving multiple stakeholders – from states to technical experts to airlines (and further downstream, businesses and passengers) – that span the globe. Nevertheless, precisely because of its heavy dependence on international governance and regulatory design, aviation serves as a particularly illustrative example to contemplate the role of institutional actors in propounding and, in many instances, frustrating equal aeromobile opportunities for all in a global context. The next section begins by reviewing recent literatures that have touched on issues of fairness and equity with respect to flying. Sections three, four and five then offer three consecutive explorations of key global institutional frameworks – namely, the regime of air rights; the regime of aeronautical expertise; and the regime of aviation security – that have contributed to making routing and the ease of travel more inimical to some nation-states than others. The concluding section reflects on the implications of these methods of governance for the world's distribution of aeromobility and argues that they are in fact significant starting points for understanding geopolitical injustices in (aero)mobility.

Mobility justice in flight

In the past decade, scholars have paid increasing attention to the unevenness of aeromobility, with certain socioeconomic groups tending to enjoy greater access to and benefits from air travel than others. While the emergence of low-cost carriers has opened up the market to larger numbers of passengers (Hirsh, 2017), there remain persistent asymmetries in who gets to travel freely, quickly and in comfort. On the one hand, it is frequently the middle class or those who are already mobile who benefit most from the liberalization of air travel (Casey, 2010). On the other, there remain sizable groups of people (i.e., the poor, undocumented and war-displaced) who have yet to even taste the socioeconomic promises of flying. Scholars often critique the injustice of this travel gap and note that the world's most disadvantaged are deprived of modernity's most prominent mobility resource by dint of their disconnection from its infrastructures (Martin, 2011). As Sheller (2010, p. 269) concurs, air transport is made up of a series of "open skies and closed gates", producing a reality that is deeply hierarchical and racialized and where, in the end, white, affluent and usually male professionals and jet-setters stand out as triumphant.

If this first set of writings foregrounds the differentiation between those who have access to aeromobility and those who do not, then a second body of work pinpoints a further politics related to the ability of some air passengers to move through the world's gateways quicker and with more ease than those relegated to economy class or a low-cost carrier (Adey, 2010). Adding a class dimension to an already racialized schism, these literatures highlight the presence of a dromologics – or a politics of speed – in the way air infrastructures are technologically arranged to sort through eligible air passengers in the name of security (Lyon, 2003). As Adey (2006, p. 198) observes in cases of global-scale mobilities, "the differentiation of movement is increasingly tied to the differentiation of people", as premium passengers and kinetic elites "experience faster and improved access", while the vast majority is slowed down to be subject to further surveillance and screening at both ends of the air journey. Focusing more specifically on the US-Canada fast-track NEXUS program, Sparke (2006, p. 168) critiques the pandering to neo-liberal ideals in global mobility, where upper-class frequent fliers are favored and enjoy "[e]xpedited airport screening ..., shorter check-in lines, valet parking, ... express lanes, and ... multiple privileges and protections" at the expense of those who have not yet attained the same class status (see also Budd & Graham, 2009, on private business jet travel). In air travel's universe, speed thus becomes politicized and unevenly rationed among different people, with the fast and the quick constituting an exclusive domain to which only elite travellers have access.

Scholarship highlighting these schisms in air travel usefully dispels any (false) notion of the ascendancy of a uniformly hypermobile world. By recounting the relativity of speed, comfort and ease in people's travels, it brings aeromobility – and other forms of movement – under the scrutiny of mobility justice. Yet, just as current research is astute in prodding the world towards a more equal distribution of mobility rights among individuals, it is relatively reticent about the larger-scale inequalities that nation-states experience in mobility and the institutional mechanisms producing them. Not to be dismissed as an analytical scale that is too abstract or distant from the practicalities of people's movement, decisions taken at the global level can have significant knock-on effects on lived mobilities on the ground. Indeed, they constitute the legal and discursive governmentalities that give substance to international norms of mobility (Salter, 2007), which in turn define the mobile relationships between entire polities and populations of people. In the same way that social behavior reflects underlying governing systems-at-large, such as the global economy (Mitchell, 2002), (some) mobilities likewise have their beginnings in these interventions that render certain kinds of international practice 'proper'. This power/knowledge enactment not only has an agenda-setting purpose; it also seeks to spell out a territorial vision of how mobilities are to be differentiated and made unequal between entire nation-states.

Using aeromobility as a springboard to apprehend the potency of these global institutional actions, the following section delineates a few fundamental tenets in aviation governance that are responsible for the geographically uneven order of international aeromobility. While these examples do not exhaust the reasons why some groups, some classes and some societies (e.g., in the Global South) find it more challenging to journey freely by air than others, they are able to provide a broad overview of the major institutional determinants that variegate different countries' access to flying. By interrogating the legal, aeronautical and security organization of the industry, I elucidate how dissimilar experiences of aeromobility can, on one level, be traced to the "protracted and convoluted processes" in global air transport planning that give macrostructural shape to aeromobility as a whole (Lin, 2018, p. 100). Understanding how these processes are the building blocks of international aviation sheds light on the foundational bases undergirding the institution of flying in modern times and helps trace new sources of injustice beyond isolated pieces of legislation that affect individuals.

International regime of air rights

Beginning at the broadest level, the world's airline network coverage is hardly uniform but patchy at best, precipitating higher travel costs for less well-connected places than those that enjoy nodal status. Consequently, global aviation in the first instance is not an evenly distributed commodity. One driver of this asymmetry is a route rationing system that was instituted at the close of World War II, known (ironically) as the Freedoms of the Air. Modeled after the Bermuda I Agreement between the United States and United Kingdom in 1946, this system has become the basis by which nation-states enter into capacity-limited exchanges of commercial air rights with one another. To be more precise, Freedoms of the Air accord each state the prerogative to determine whether or not, and in what form, another country's airline(s) can have the "privilege" to "enter and land in [its] airspace" (Hinkelman, 2008, p. 81). As Raguraman (1986, p. 66) highlights, the "regulation of the supply of international air transport services through governmental control over capacity and route rights [directly affects] air traffic flows, networks and airline operations". Although these exchanges have been liberalized over time, "the broad structure" of this framework, which requires states to negotiate bilateral and multilateral treaties between themselves before commercial air services can be launched, remains firmly "in place" (Button & Taylor, 2000, p. 211).

This permission-based system of bartering routes might appear merely procedural, but it is in fact extremely political. Whereas large affluent countries in the Global North tend to maintain an upper hand in shaping these exchanges, states in the Global South, particularly those lacking the financial wherewithal to develop their aviation industries, tend to receive fewer rights in return for their own airlines (Raguraman, 1986). This imbalance in bargaining power results in lopsided exchanges of air rights where stronger states are able to secure better access for their network carriers and, in the process, capture greater capacity and market share. At the same time, by curtailing the growth of foreign airline entrants in their territories, they are able to stay shielded from usually cheaper airlines from the Global South. The barter between the UK and the newly independent Singapore in the 1960s illustrates the structural asymmetries sanctioned by the Freedoms of the Air legal framework. In that episode, Britain had refused Singapore's carrier the right to fly to London. As a Singaporean negotiator later recounted, British counterparts held the belief that "Singapore is nothing [while] London is so valuable", claiming that what Singapore was offering Britain – the right for British carriers to fly to Singapore – "is not good enough" for Singapore's airlines "to get to London" (Lim, 2000). The case was eventually resolved through risky political brinkmanship; British carriers' colonial rights to fly to the city were altogether withdrawn, and Singapore isolated itself from London. It potently highlights the difficulties that smaller states face in becoming fully aeromobile due to the protectionist stance allowed in international air law.

This Global North strategy of gaining a monopoly in airline routes through unequal endowments in air rights has made its appearance in the Global South too. As a rising power willing to expand its aerial network and to afford its people more reliable service through its national carriers, China too has recently sought to stymie the entry of foreign carriers through the air rights system. With the aim of augmenting the viability of nascent domestic carriers that needed time to expand, the Civil Aviation Administration of China (CAAC) has adopted a protectionist stance against international competitors such as Gulf and US carriers to prevent them from competing with Chinese airlines in the home market (Center for Aviation, 2013). In 2017, the CAAC rejected American Airlines' application for landing slots on a Beijing-Los Angeles route over which Air China had a monopoly. This protectionist move prompted American Airlines to lobby its government to block CAAC's future requests for air rights, in order to protect US airline development from Chinese expansion (Center for Aviation, 2017). Such rivalries signal that the battle for air rights (and routes) – and, by extension, a population's ability to gain access to a robust set of autonomously controlled air services – is fought at the international level. Before aeromobility is possible, an institutional win in the contest for Freedoms of the Air must first be secured.

The intergovernmental practice of air rights exchanges – and the imbalanced patchwork of airline networks resulting from it – may seem at first to be tangential to individuals' ability "to exercise rights to mobility" (Sheller, 2013, p. 186). However, it is the foundational basis of (aero)mobility justice given the way it limits a nation's access to airline resources. On the international front, this legal framework constitutes the ground rules for the world's allocation of routes. Rather than being evenly distributed, these rules, via the mechanism of air rights, are often politicized to the end that select states and their carriers might dominate global thoroughfares (Aaltola, 2005). For citizens of these countries, the concentration of airline routes in their jurisdiction enables them to partake in aeromobility more easily than those living in less influential countries, who face fewer choices, more inconvenient connections and higher fares. Without a viable airline network supported by a favorable suite of air rights, aeromobility may still prove a distant dream, and an unjustly foreclosed opportunity, for this latter group of people.

International regime of aeronautical expertise

Legal barriers to the world's air routes are only the beginning of the uneven distribution of aeromobility among nations. Even after air rights are procured, authorities are still required to adhere to global aeronautical regulations, which set capacity limits on a country's or region's ability to mount flights. Bound by international agreement, states are expected to follow the directives of the (US-led) International Civil Aviation Organization (ICAO), whose mission is "to secure international co-operation [and the] highest possible degree of uniformity in [navigational] regulations" (ICAO, no date/a). Because many countries did not exist when the ICAO was founded, the first rules of the air were written by 'traditional' air powers in an inevitably skewed exercise.

In March 1946, the United States, Canada, Britain and France gathered in Dublin to arbitrate a pilot suite of procedures for transoceanic travel at the International Conference on the North Atlantic Route Service Organization. To cater to the 'unique' airspace environment of Europe (typified by short-haul, overland flights), a second version of these procedures was formulated two months later in Paris at a parallel conference for European-Mediterranean states (ICAO, 2007). These rules were then made "applicable on a worldwide basis" (ICAO, 2007, p. vii) at the close of the 1940s, with no special provisions made for distinct regions. Instead, air navigation plans for the rest of the world – including the colonial regions of the African Indian Ocean, Caribbean, Middle East and Southeast Asia – were discussed at Regional Air Navigation Meetings (RANMs) held between 1946 and 1953 (MacKenzie, 2010). Rather than experimenting with new practices, the goal of these meetings was to assess these regions' facilities and determine if additional equipment was needed to bring them on par with the Dublin and Paris frameworks.

This diffusion of technical standards from the North Atlantic and Mediterranean regions to the rest of the world may seem like a form of technical assistance that pioneering states were extending to their more 'backward' counterparts (Caprotti, 2011). However, aeronautical rules formalized according to technical principles familiar to Western European and North American governments and aircraft manufacturers had far-reaching implications for the control of aeronautical expertise from the 1940s onward. Indeed, after seven decades, these dominant states' stronghold over decision-making processes related to air navigational techniques and airspace capacity increases remains intact. This situation can partly be explained by the technological edge dominant countries maintain over others; but equally crucial is the persistent concentration of 'legitimate' knowledge about aircraft, avionics and airspace design in the hands of a few Global North powers. Without the ability to upend this trend, the rest of the world will not be able to rewrite and share equally in aviation's aeronautical order.

This technological domination subtly distorts aviation development among nations by causing air traffic capacities to grow at different speeds in different places. Non-leading states cannot advance alternative programs that would improve their own aeronautical capacity and related socioeconomic potential on their own terms. Instead, to be accepted into the international community of 'safe' airfaring nations, countries in Asia, Africa and Latin America are compelled to follow Western technology and to cede their aeromobility potentials – of possibly greater traffic volumes, greater speeds, and greater airspace densities through indigenous aeronautical innovations – to the leaders in Western Europe and North America. Take the example of Reduced Vertical Separation Minima (RVSM). A landmark initiative that doubles airspace capacity by halving the required vertical separation between aircraft, the program was first implemented in North Atlantic airspace in 1997. Only a decade later was it diffused to the rest of the world, with guidance from the US Federal Aviation Administration (ICAO, 2012). Or consider the spread of the aircraft-tracking technology known as Automatic Dependent Surveillance-Broadcast. In 2001, it was introduced to improve air traffic efficiency and safety in Alaska. Like RVSM, it was also incorporated in the Asia-Pacific, Caribbean and African regions only after a period of time had elapsed (ICAO, 2016). These examples demonstrate how the Global South is technologically dominated by 'traditional' air powers and entrapped in a relentless cycle of (waiting for) unidirectional expertise transfers from the North. Augmenting aeromobility, therefore, does not consist of achieving growth potentials autonomously, but rather trying to close a perpetual technological gap with the Global North.

The divergence in aeronautical know-how between unequally positioned states in the world's technological hierarchy constitutes another instance of (aero)mobility injustice. A lack of state-of-the-art equipment or expertise is not the issue. Rather, as Global South countries lag behind preeminent states in terms of aeronautical capabilities, they will be left relatively constrained in growing air traffic until a technical transfer takes place. In particular, the iterative loop of being bound by 'international' aeronautical rules, which in turn require 'international' assistance to augment aeronautical efficiency, solidifies a situation of technological dependency (Gereffi, 1983). It also contains aviation development among 'follower' states and keeps it perennially one step behind their stronger peers. An airspace planner in Asia captures the gravity of this situation:

What happens if you lose that connectivity to the rest of the world? People will shun you. Do they still want to come here and do business? Do they want to invest? It's not that they won't. It becomes difficult.

(Personal communication)

The Global South faces this challenge of maintaining a competitive advantage in connectivity when it cannot outcompete the North in aeronautical capacity expansion. This power imbalance, too, deserves attention in order to help 'lagging' states break free from their current technological dependency and attain a future of aeromobility autonomy.

International regime of aviation security

Previous sections have detailed how Global South countries are prone to international restrictions on their airline networks and technical maneuverability as instances of (aero)mobility injustice. Here I interrogate the institutional mechanisms that govern international travel in the name of security. Oftentimes, scholars have approached aviation security as a set of border processing measures that discrete governments take to sort and differentiate the various kinds of travellers crossing into their territories (Adey, 2010; Salter, 2007; Sparke, 2006). However, the injustice of these variegated mobilities emerges not only at the border; it has its genesis in wider structural frameworks at the inter-state level. In this instance, Global North countries, again, exert an inordinate amount of influence on traveller processing practices, also known as passenger facilitation, worldwide. These interventions constitute the international guidance on how passengers are to be accepted for air travel in every country (ICAO, 1980). More profoundly, this seemingly innocuous act of institutionalizing passenger facilitation procedures sets the tone for how nations deal with each others' citizens, as well as their air passengers, at a time of increased mobility.

Central to this global guidance on passenger facilitation is the (state-based) practice of issuing passports to citizens who undertake international travel. The passport regime was introduced in the 19th and early 20th centuries to selectively control the flow of labor, criminals, refugees and aliens across colonial territories. However, it did not gain traction globally until it was adopted by the ICAO as an integral part of air travel (Salter, 2003). In 1968, the ICAO tasked a Panel on Passport Cards (POPC), working under the auspices of the Air Transport Committee, to develop a standardized travel booklet or card that would clear passengers through passport controls more quickly. In the ensuing decade, the panel, headed by Australia, Canada and the US, recommended adopting a new generation of globally inter-operative passports. In 1980, ICAO Document 9303 formalized and promulgated these recommendations for global implementation and consensus. With inputs from the same leading states, it further advocated developing "a passport with machine readable capability" using the optical character recognition technology (ICAO, 1980). Based on these recommendations, the United States, one of the panel members, issued the first machine-readable passports (MRPs) in 1981, which became a global prototype (Lyons, 1981).

These institutional developments initiated a chain of events that produced a (familiar) global convergence in passport design and security features based on US standards. In 1984, Americans established the Technical Advisory Group on Machine Readable Travel Documents (TAG/MRTD) "to advise States and review as well as update passport specifications" (Heitmeyer, 2010, p. 2). The group comprised government officials from the United States and ten other ICAO member states that were deemed to be experts in border controls (including Australia and Canada as well as France, Germany, Japan, the Netherlands and the United Kingdom). It inherited the mandate of the POPC, but also expanded its work to experiment with new technologies and assist in passport implementation, education, promotion and capacity-building among states. At the end of TAG/MRTD's work to ensure that all states adopt the standard machine-readable passport, the ICAO conducted security audits on all member states, aiming to enhance aviation security, identify performance deficiencies and ensure uniformity in security-related standards, including that of passports (ICAO, no date/c). By laying down stringent parameters for the documentary aspect of passenger facilitation, the ICAO in effect became a US vehicle for institutionalizing a particular model of 'secure' aeromobility, acting concomitantly as an institutional instrument to correct noncompliant states.

These passport stipulations critically altered the way people of different nationalities moved. After MRPs were introduced, citizens of countries that promptly complied with the new passport standards – the United Kingdom, Germany and numerous other Western European states (Dodsworth, 1986) – were able to benefit from accelerated border processing (especially to the United States) on the alleged grounds that they held more secure, interoperable and fraud-free documents; citizens of other countries, in contrast, faced increased inspection and were excluded from participating in visa-waiver programs. Besides the uneven pace of MRP implementation, deviations from the (US-modeled) standard MRPs were also a problem for some countries. Alphanumeric machine-readable zones (MRZ) suited the written scripts used in North American and European passports. But they were incompatible with "Arabic, African and Asian name formats, the use of scripts other than Roman, and the [presence] of diacritical marks in some languages" (Chatwin, 2011, p. 2). Moreover, some countries used different passport laminates and typesets which standard readers could not recognize. These peculiarities created a speed-differentiated aeromobility regime where populations from the Global North were able to travel more freely than other national groups, which were slowed down, or even immobilized, due to their 'substandard' passports.

New passport technologies, which incorporated biometrics and data storage capabilities in passports, were implemented in the aftermath of September 11th, accentuating these differentiations. Within weeks of the attacks, the United States decreed that all states wishing to remain on its Visa Waiver

Program would have to meet biometric standards promulgated in the ICAO's revised Document 9303 (Chatwin, 2011). While some countries such as Japan, Singapore and the United Kingdom quickly issued biometric documents to meet these requirements, the majority of states held off due to concerns about cost, technology and privacy, effectively eroding their citizens' ease of travel within the new aeromobility order. The ICAO set up a separate public key directory (PKD) in 2007 that provided states issuing biometric passports a platform to share their documents' unique digital certificates – signatures embedded in the biometric data for authentication purposes – on a secure database, allowing their immigration officers to cross-validate each other's passports at international borders. Each state needs to pay US$15,900 to register into the system and over US$32,500 per annum in operating fees. Consequently, the PKD remains a relatively undersubscribed, if exclusive, program, with only 58 participants as of 2017 (ICAO, no date/b). Implementing these security measures at such an uneven pace ensures the unequal distribution of (aero)mobility opportunities – and mobility justice – among citizens of different countries. As the Global North defines the parameters of 'secure' aeromobile travel, it renders citizens of much of the Global South – especially from countries like Afghanistan, Algeria, Kiribati and Mauritania that did not even have MRPs by 2011 (ICAO, 2011) – less fit and able to fly.

Conclusions

This chapter has outlined the workings of three international regimes related to aeromobility and explained how they contribute in less conventional ways to our understanding of the relationship between mobility justice and aeromobility. It supplements existing literatures that delineate the politics of movement among air passengers (Adey, 2006) by pointing to an institutional and macrostructural form of discrimination in aerial movement. Indeed, considerations of fairness or equity in one's ability "to exercise rights to mobility" (Sheller, 2013, p. 186) entail more than inequalities structured along social axes of race, class and gender. Injustices also arise in the systematic way 'global rules' splinter the mobility rights of whole populations in different nation-states. By interrogating the tripartite regimes of air rights, aeronautical expertise and aviation security, I demonstrated that asymmetric privileges in the organization of aviation can leave lasting and far-reaching impacts on the mobilities of entire jurisdictions of people, in much the same way as race, class and gender do on an individual basis. Like an upscaling of the social injustices experienced by marginalized persons and groups, these aeromobility exclusions speak of the marooning of *places*, especially in the Global South.

Moving to this scale of analysis contributes two valuable insights to the current analytical toolbox of mobility justice. The first pertains to the need to pay attention to the global institutional structures that determine the spatial outcomes of movement at the international level. These global governance structures seldom refer to isolated pieces of legislation, but invoke the collectiveness of a diffuse network of (unequally empowered) actors, decision makers and agencies that play a part in organizing mobilities around the world (Lin, 2018). Because these institutional arrangements are hierarchical, they also usually enact an international order which not only defines the relations of mobility between states, but also constrains alternative imaginings of those relations. Oftentimes, such methods of rule involve monopolizing expert knowledge in strategic fields (Mitchell, 2002), which legitimizes particular ways of understanding that foreclose a more just and equitable sharing of mobility's benefits across places. As I have shown, this is exactly the maneuver that 'traditional' air powers have used to gain comparative advantage for themselves, in the process naturalizing particular self-serving rationales related to the legality of routes, the control of aeronautical expertise and the appropriateness of security measures that curtail other polities' ability to be aeromobile.

A second focal shift involves attending to the intersections between mobility justice and geopolitics. While there is no dearth of research explicating the *politics* of mobility (or the manners in which mobility experiences are splintered among individuals), the equally important ways in which states and state actors secure mobility rights geostrategically to achieve larger political goals of economic development and international domination remain largely unexplored. Mobility calculations of this sort are not new; they have long shaped the way institutional actors – from colonial governments to modern states – envision globally mobile worlds (Caprotti, 2011; Raguraman, 1986). Insofar as these geopolitical strategies impact how a nation's citizens can move, there is ample scope for advocates of mobility justice to seriously contemplate the effects of and the attenuating responses to such high-level interventions. Although the activities of global institutions are often obscure, rendering them less amenable to mobility-informed analyses, these nodes of governance may be where the structural bases of unequal and unjust mobilities are ultimately found.

References

Aaltola, M. (2005). 'The international airport: The hub-and-spoke pedagogy of the American empire'. *Global Networks*, 5(3), 261–278.

Adey, P. (2010). *Aerial life: Spaces, mobilities, affects*. Malden, MA: Wiley-Blackwell.

Adey, P. (2006). '"Divided we move": The dromologics of airport security and surveillance'. In T. Monahan (Ed.,) *Surveillance and security: Technological politics and power in everyday life* (pp. 195–208). London: Routledge.

Adey, P., Budd, L. & Hubbard, P. (2007). 'Flying lessons: Exploring the social and cultural geographies of global air travel'. *Progress in Human Geography*, 31(6), 773–791.

Budd, L. & Graham, B. (2009). 'Unintended trajectories: Liberalization and the geographies of private business flight'. *Journal of Transport Geography*, 17(4), 285–292.

Button, K. & Taylor, S. (2000). 'International air transportation and economic development'. *Journal of Air Transport Management*, 6(4), 209–222.

Caprotti, F. (2011). 'Visuality, hybridity, and colonialism: Imagining Ethiopia through colonial aviation, 1935–1940'. *Annals of the Association of American Geographers*, 101(2), 380–403.

Casey, M. E. (2010). 'Low cost air travel: Welcome aboard?' *Tourist Studies*, 10(2), 175–191.

CenterforAviation. (2017). 'US–China open skies: A window in 2019 – alignment of airline partnerships & airport infrastructure'. *Center for Aviation*. Retrieved from https://centreforaviation.com/insights/analysis/us-china-open-skies-windo w-in-2019-with-alignment-of-airline-partnerships-airport-infrastructure-340603.

CenterforAviation. (2013). 'Gulf carriers and Turkish Airlines ready to expand in China'. *Center for Aviation*. Retrieved from https://centreforaviation.com/insights/a nalysis/gulf-carriers-and-turkish-airlines-ready-to-expand-in-china-if-only-air-rights-were-available-136090.

Chatwin, C. (2011). 'The story of standardization: A history of ICAO and ICAO Document 9303'. *Keesing Journal of Documents & Identity*, 36, 1–6.

Cwerner, S. (2009). 'Introducing aeromobilities'. In S. Cwerner, S. Kesselring & J. Urry (Eds.), *Aeromobilities* (pp. 1–22). London: Routledge.

Dodsworth, T. (1986, December 3). 'Machine-readable passport planned'. *The Financial Times*, p. 20.

Gereffi, G. (1983). *The pharmaceutical industry and dependency in the third world*. Princeton, NJ: Princeton University Press.

Heitmeyer, R. (2010). 'ICAO civil aviation and MRTD standards'. *Keesing Journal of Documents & Identity*, 31, 1–6.

Hinkelman, E. G. (2008). *Dictionary of international trade: Handbook of the global trade community*. Petaluma, CA: World Trade Press.

Hirsh, M. (2017). 'Emerging infrastructures of low-cost aviation in Southeast Asia'. *Mobilities*, 12(2), 259–276.

ICAO. (2016). 'Enhancing safety and expanding capacity: Implementation of ADS-B out in the United States'. Retrieved from www.icao.int/Meetings/a39/Documents/ WP/wp_174_en.pdf.

ICAO. (2012). 'Asia/Pacific Regional RVSM monitoring statement'. Retrieved from www.icao.int/APAC/Documents/edocs/Asia%20Pacific%20RVSM%20Monitoring% 20Statement%20V%201.pdf.

ICAO. (2011). 'Current status of states in relation to the implementation of Document 9303'. Retrieved from www.icao.int/Meetings/TAG-MRTD/Documents/Tag-Mrtd-20/ TagMrtd-20_WP010_en.pdf.

ICAO. (2007). *Procedure for air navigation services: Air traffic management*, Doc. 4444. Montreal, QC: International Civil Aviation Organization.

ICAO. (1980). *A passport with machine readable capability*. Montreal, QC: International Civil Aviation Organization.

ICAO. (no date/a). 'Foundation of the International Civil Aviation Organization'. Retrieved from www.icao.int/about-icao/pages/foundation-of-icao.aspx.

ICAO. (no date/b). 'ICAO PKD'. Retrieved from www.icao.int/Security/FAL/PKD/Pa ges/default.aspx.

ICAO. (no date/c). 'The universal security audit program continuous monitoring approach (USAP-CMA) and its objective'. Retrieved from www.icao.int/Security/ USAP/Pages/default.aspx.

Lim, C. B. (2000). Oral history interview, September 5, Accession Number 002358/03 [Transcribed by K. Koh]. Singapore: National Archives.

Lin, W. (2018). 'Transport provision and the practice of mobilities production'. *Progress in Human Geography*, 42(1), 92–111.

Lyon, D. (2003). 'Airports as data filters: Converging surveillance systems after September 11th'. *Journal of Information, Communication & Ethics in Society*, 1(1), 13–20.

Lyons, R. D. (1981, April 19). 'State dept plans to introduce new machine-readable passport'. *The New York Times*, p. 19.

MacKenzie, D. (2010). *ICAO: A history of the international civil aviation organization*. Toronto: University of Toronto Press.

Martin, C. (2011). 'Desperate passage: Violent mobilities and the politics of discomfort'. *Journal of Transport Geography*, 19(5), 1046–1052.

Mitchell, T. (2002). *Rule of experts: Egypt, techno-politics, modernity*. Berkeley, CA: University of California Press.

Raguraman, K. (1986). 'Capacity and route regulation in international scheduled air transportation: A case study of Singapore'. *Singapore Journal of Tropical Geography*, 7(1), 53–67.

Salter, M. B. (2007). 'Governmentalities of an airport: Heterotopia and confession'. *International Political Sociology*, 1(1), 49–66.

Salter, M. B. (2003). *Rights of passage: The passport in international relations*. London: Lynne Rienner Publishers.

Sheller, M. (2013). 'The islanding effect: Post-disaster mobility systems and humanitarian logistics in Haiti'. *Cultural Geographies*, 20(2), 185–204.

Sheller, M. (2010). 'Air mobilities on the US–Caribbean border: Open skies and closed gates'. *The Communication Review*, 13(4), 269–288.

Sparke, M. B. (2006). 'A neoliberal nexus: Economy, security and the biopolitics of citizenship on the border'. *Political Geography*, 25(2), 151–180.

4 Fleeing Syria – border crossing and struggles for migrant justice

Suzan Ilcan

Introduction

Social scientists – in concert with journalists, photographers, bloggers and humanitarian personnel – frequently represent border-crossing migrants, refugees and asylum seekers in sensational and contradictory ways, either as victims of brutal bordering practices and policies or as unruly and disorderly mobile subjects who pose threats to national security. Challenging the view of migrants as victims (Bargu, 2017; Hess, 2017) and unruly subjects (Cresswell, 2006; Isleyen, 2018), I focus on Syrians who fled Syria between 2011 and 2017, their encounters with unjust bordering practices during these journeys and the explicit demands for migrant justice these encounters cultivated. I detail how Syrian migrants negotiate border practices through struggles for migrant justice and what implications these struggles have for mobility justice.

To address these questions that connect bordering practices and mobile groups, I draw on and contribute to critical migration and border studies literature (e.g., Bargu, 2017; Hess, 2017; Isleyen, 2018; Johnson, 2014; Mountz, 2011) and the literature linking mobilities and justice (e.g., Cook & Butz, 2016; Ilcan, 2013a; Rygiel, 2013; Sheller, 2014). I argue that in the attempt to control migrant movements, bordering practices also nurture calls for migrant justice. 'Migrant justice' refers to demands for justice by those on the move who seek access to freedom of movement, safety, protection, rights or citizenship. In the Syrian case, it invokes the political agency of migrants and refugees in making the decision to move, in negotiating bordering practices and in mobilizing their membership for inclusion or recognition in host states. Migrant justice is not merely a means to achieve political ends, but is itself a power relation and way of challenging border and state authority during migrant journeys. The struggle for migrant justice is an element of 'mobility justice', an umbrella concept used to conceive the relations of power that shape the movements of people, resources and information at local, regional, national and international scales (Cook & Butz, 2016; Sheller, 2014, 2018).

My analysis draws on policy and program documents and 53 semi-structured interviews with Syrian refugees that foreground their struggles to negotiate militarized checkpoints and territorial borders and access the safety and protection of host states. The interviews, which consist in part of Syrians' retrospective narrating of their journeys out of Syria to nearby host states, were conducted in Arabic or English, and they lasted 90–120 minutes. They took place in Kitchener, Mississauga and Toronto, Canada, between January and December 2017. Participants included 24 women and 29 men who lived in Syria both prior to and during the civil war. The majority self-identified as Syrian; the remainder as Kurdish, Palestinian or Turkmen.

The chapter is comprised of three sections. The first develops the background for understanding how border control practices regulate migrants while simultaneously fostering their struggles for migrant justice. The second provides the context for understanding Syrians' journeys in relation to the war. The third section analyzes Syrians' struggles for migrant justice as they flee to nearby host states. In their flight from the vulnerabilities of war, migrants make use of unconventional methods and intermediaries to thwart unjust bordering practices. I conclude by showing how struggles for migrant justice provide insights into the broader theme of mobility justice. My use of 'mobility justice' highlights the politics of migration and border management as key dimensions of the struggles migrants, refugees and asylum seekers confront in their everyday lives. My analysis of migrant border-crossers, therefore, shifts them out of the domain of victimhood and into the domain of political agency in the context of struggles for migrant and mobility justice.

Border-crossing and struggles for migrant justice

An intensified interest in the mobility of people, governing practices and political relations has prompted a 'mobility turn' in the social sciences (e.g., Adey, 2016; Cook & Butz, 2016; Cresswell, 2006; Ilcan, 2013a; Mountz, 2011; Sheller, 2017; Urry, 2007). This turn has led to a focus on new relations of power and visions of justice that avoid established disciplinary boundaries and sedentarist understandings of the social (Ilcan, 2013b). The mobilities paradigm, in part, is informed by research that explores what happens 'on the move' as a way to understand how mobilities are produced (Cresswell, 2006; Stasiulis, 2013), controlled (Brodie, 2013; Rygiel, 2013) and contested (Ilcan, 2013c; Mainwaring & Brigden, 2016). It overlaps with many thematic areas, such as globalization, tourism and border studies, and has methodological implications for studying migration by reconceptualizing migration as an ongoing process rather than a discrete event (Schapendonk & Steel, 2014). Consequently, it has led to new understandings of migration as a journey produced on the move, but not necessarily bounded by separate beginnings and ends (Collyer, 2010; Cranstron, Schapendonk & Spaan, 2018) or devoid of border contestations (Genç, Heck & Hess, 2018). Together with work in critical migration and border studies (e.g., Hess, 2017; Squire, 2014), the mobilities literature highlights the (im)mobilities migrants face due to massive conflicts and their legislative, policy and socioeconomic dynamics (Innes, 2015; Isleyen, 2018; Stasiulis, 2013). For example, refugees fleeing conflict often experience long periods of waiting, hiding, stopping and containment (Doty, 2011), insufficient rights and state policing (Üstübici, 2016) at border crossings.

Bordering practices are proliferating to immobilize some groups while mobilizing others, such as those experiencing war, persecution and insecurity. They are diverse, and include militarized border crossings (Nevins, 2002), border and regional policies (Genç, Heck & Hess, 2018; Pickering & Weber, 2006) and the formation of new policing and security measures that regulate certain types of movement, in the process creating precarious living conditions for migrants and refugees (Ilcan, Rygiel & Baban, 2018; Isleyen, 2018). Bordering practices have culminated in the social sorting and separating of mobilities (Walters, 2006), the labeling of migrants and refugees as both at risk and a risk (Pallister-Wilkins, 2015) and the shaping of mass movements of peoples along gender, ethnicity and class-based lines (Gerard & Pickering, 2014), as well as in increases in migrant deaths, particularly along the US–Mexico border (Carlson & Gallagher, 2015; Squire, 2014). Furthermore, amplified border and migration management efforts, such as along the borders of Jordan, Lebanon and Turkey, have controlled Syrian refugee movements since the start of the 2011 war. Such bordering practices are frequently part of state policies, or what Mountz (2011, p. 323) refers to as 'state mobilities', that involve suppressing people on the move through border, visa and other control initiatives.

Bordering practices are more than mechanisms for othering migrants and controlling their movements, as they can also foster struggles for migrant justice. These struggles involve migrants and refugees negotiating bordering practices as part of an effort to move or escape, to stay in one country and not another and to exercise the rights granted to them as migrants or refugees (Ataç et al., 2017; Tyler & Marciniak, 2013). They are exemplified in migratory journeys and stories (e.g., Almustafa, 2018; Collyer, 2010; Genç, Heck & Hess, 2018; Mainwaring & Brigden, 2016) which, as I discuss below, can also highlight the broader issues of mobility justice.

In emphasizing the political landscape of the Syrian conflict that prompted many Syrians to flee the country, the analysis below centers on those who decided to leave and negotiated bordering practices during their journeys. Such negotiations enabled them to disconnect from the brutal injustices and vulnerabilities of war and provided them with temporary access to safety and protection in nearby host countries.

"And planes are bombing kids": fleeing Syria during the war

In March 2011, the Syrian uprising began with protests demanding regime change and political reform, which gradually transformed into a brutal conflict that by the end of 2015 internally displaced more than 7 million Syrians and forced more than 4 million to leave Syria for neighboring Lebanon, Jordan, Turkey, Iraq and Egypt (Thorleifsson, 2016). Based on my interviews with Syrian refugees, many of them experienced the effects of war and bordering practices in the Syrian towns and cities where they lived, such as Aleppo, Daraa and Homs. Many recall the public fears of the war – homes, schools and neighborhoods bombed, and citizens tortured, raped or killed – and peoples' political responses to military and state power that made their bodies sites of oppression, demonstrating that "the border" can be "everywhere that an undesirable is identified" (Agier, 2011, p. 50), including the indeterminate zone in which the traveler's body *becomes* the border, the site of persecution or oppression.

Inspired by Egyptian and Tunisian protesters, 15 school children in the Syrian provincial town of Daraa, located 13 kilometers north of the border with Jordan, engaged in antigovernment activities by writing slogans of revolution on the walls of their school, such as: "The people want to bring down the regime". They were soon arrested by the Moukhabaret (intelligence service) and tortured. In response to the state's brutal treatment of them, many people in Daraa took to the streets to engage in 'peaceful protests'. During and following these protests, demands for justice were mobilized (interviews with Syrian refugees, Kitchener, Mississauga and Toronto, 2017) through tweets, Internet posts, digital media recordings and videos that circulated within and outside Syria (see Urry, 2007, on "virtual travel"). Such mobilizations contributed to mass demonstrations that extended beyond the weekly rhythm connected with Friday prayers to become daily events (Leenders & Heydemann, 2012, p. 149).

The state's continued crackdown on protesters – including mass shootings by security personnel and the systemic torture of people arrested for taking part in antigovernment activities – made the waning of basic political freedoms a key public concern for countless Syrians. In my interviews, many Syrians recalled the regime's violent responses to the protests, the relentless bombings and the military apprehension of some children on city streets and at public schools and universities. Streets and schools thus became diverse sites of state bordering practices or of reterritorialization within Syria, reflecting what Weber and Pickering (2011, p. 13) refer to as the "complex performances of state power staged at multiple locations". Many Syrians acknowledged the authoritarianism of President Bashar al-Assad's regime and the inability of the Syrian state to provide secure lives and protection for its nationals. For example, one male Syrian living in Toronto stated that he experienced the necessity

for freedom from the regime, the state of emergency, curfews, the prison, silencing people and the media control by the regime and its allies of monopolies in the industries (cars, food, medicine). The regime and only a small number of people control all of the economy.

He stressed the need for

human dignity, the dignity as a right to live as a citizen. Not a single soul there feels dignity...Social equality was nonexistent. In Syria, there is no such thing. What equality? And planes are bombing kids. Syria lacks citizenship laws.

Syrians' insufficient access to safety, protection and citizenship rights impelled them to flee for nearby countries and to confront bordering practices during their journeys.

Negotiating border crossings

Military checkpoints

During their journeys, Syrians encountered different bordering practices throughout Syria, including blocked or semi-blocked lanes, regulated neighborhood passage routes and militarized checkpoints at bus stations and schools and on roadways where interrogations, abductions, arrests and sanctioned crossings occur. These sites of reterritorialization and bordering within the state, which are often uneven and far from static (see Little, 2015, p. 436), can involve people on the move responding to the injustices of reterritorialization, such as through struggles for migrant justice that often involve knowledge of the politics of border-crossing and the ability to negotiate bordering practices in an effort to move and access just living conditions, as I discuss below.

In interviews, Syrians spoke about their journeys to Syrian–Turkish, Syrian–Jordanian and Syrian–Lebanese borders, and their (legal and illegal) capacity to enter military checkpoints under conditions of heavy securitization and to negotiate these crossings. Passing through military checkpoints often involves border-crossers showing identification, being subject to security interrogations (and occasionally violence) and challenging the technical and bureaucratic aspects of bordering that aim to regulate their movements. For example, one family that currently resides in Mississauga decided in 2015 to flee their home city of Hajar al Aswad, 4 kilometers south of the center of Damascus, and head to Lebanon, a country that currently hosts over 1.2 million Syrian refugees (Lambert, 2017, p. 738). Recalling the journey, they emphasized their capacity to challenge bordering practices. A 60-year-old man described his ability to successfully negotiate his family's way through Syrian military checkpoints:

> When we came near the [Syrian-Lebanese] border, there were also mass groups of people, a sea of them. To get there, we went through an area that was controlled by the regime. They stopped us and searched us. We felt like we were going to die. Then we died [emotionally]. They ask for your name and take your ID. Then they make you stand in a straight line. … We didn't say we are going to Lebanon. We told them we are going to Mazeh. We told them we are going to this area controlled by them [the regime]. We also told them the name of [his wife's relative] who was well known and worked with the government.

Like other border-crossers, this family articulated their knowledge of the politics of border-crossing, the violent practices that can occur at border sites and their political motivation to access just living conditions. Their ability to negotiate bordering crossings highlights their political agency in making decisions to escape the injustices of war and to access protection in host states, even if it was limited.

Other migrants and refugees negotiated militarized checkpoints in ways that made visible the connection between the personal security attached to border-crossers and the military security attached to border sites. For example, one woman who was interrogated by military personnel at a checkpoint on her way to the Syrian-Lebanese border recalled being asked why her six children accompanied her and why they were not staying in the country. Before responding, she remembered advice from family members who had facilitated their residency in Lebanon to avoid eye contact with security personnel and remain silent as ways to safeguard her security. Her response was silence: "I said nothing [to the military personnel], not even a word. They let me and my children go". Such acts of silence do not indicate that migrants are rendered voiceless or lack agency or rights. Rather, in a struggle for migrant justice, refusals to speak can enable them to negotiate interrogation circumstances and potential immobilities at checkpoints and to escape unjust living conditions and access safety and protection in host states. Silence in the context of struggles for migrant justice shares similarities with other refusals to speak, like hunger-striking protesters who withhold speech "under conditions of injustice in which a shared sense of humanity has become impossible" (Bargu, 2017, p. 13). By refusing to speak, protesters publicly reject belonging to a "common *community of speech*" (Bargu, 2017, p. 13).

In the Syrian border-crossing context, silence may have unpredictable effects, but nevertheless it may enable those on the move to pass safely though checkpoints under highly securitized and threatening conditions. These and similar strategies, such as improvising one's movements under fluctuating circumstances or enlisting the aid of intermediaries during migratory journeys, open possibilities for rethinking migrants' and refugees' political actions. As responses to power relations and spaces of conflict, these actions are integral to the desire for social change and social justice and convey the political agency of those on the move. In the face of social injustices and inequalities, such political agency can foster the capacity to (re)construct the social world and recreate collective ways of participating in the world that challenge relations of power and advance more just social practices. These agency-inspired transformations are a critical dimension of mobility justice.

Territorial borders

Some Syrians left their homes for the territorial borders of Jordan, Lebanon and Turkey without formal identity papers, which could subject them to state interrogations and long-term or temporary immobilities that prevent 'illegal' border crossings. Territorial bordering practices are sites of state control of migration, which can manifest in uneven, subjective and discretionary ways (Ataç et al., 2017; Isleyen, 2018; Williams, 2015). They are also sites for migrants, refugees and asylum seekers to pursue safety and protection in host states.

Some migrants recollected the immobilities they experienced during their journeys to the territorial borders of Jordan, Lebanon and Turkey, including long periods of waiting, stopping and containment. A good example is the temporary closure of the Syrian-Lebanese border, which has occurred on and off since 2013. Some Syrians I interviewed view the closure as a way for the military to govern the border, subject them to social injustices, control access to humanitarian assistance and compel border-crossers to wait in an over-crowded space for long periods of time without knowing when the border will reopen or if they will be able to pass (Ilcan, 2017). This waiting space became a site for containing and managing the physical existence of regular and irregular border-crossers, who are often viewed as risky subjects by Syrian and Lebanese states due to their perceived foreignness, potential for criminal behavior and lack of residence or citizenship status (see Isleyen, 2018, on irregular crossings). Instead of returning to their homes or neighborhoods in Syria in the face of persistent massive violence at and beyond border crossings, migrants and refugees were determined to reach safety and to acquire some form of protection in Lebanon as residents, workers or UNHCR-registered refugees. So they stayed at the border site with their families and friends, set up tents and shared resources (food, water, information) with others until the border reopened and they could continue their journey to Lebanon. Even in the face of coercive state and border mobilities, Syrians' struggles for migrant justice took the form of seeking the ability to move and leave and to access a relatively safer and more protected environment.

Until 2015, border-crossers between Syria and Lebanon were fairly unrestricted, which is largely attributable to the long-established policy of not requiring visas at the border. However, with the increasing number of Syrians entering Lebanon from late 2015 onwards, the state started to restrict their movements by creating stricter visa and work requirements, which in turn made state mobilities visible at border sites. For example, in April 2014 the Lebanese state shut down 18 unofficial border-crossing points which it had previously tolerated, and by October 2014 it only accepted refugees in "exceptional" or "urgent" humanitarian circumstances (Turner, 2015, p. 390). In 2015 the Lebanese state also changed its residency requirements for Syrians, which made it legally necessary to have a UNHCR registration certificate to reside in the country (Janmyr & Mourad, 2018, p. 2).

In negotiating these territorial bordering practices, many Syrians I interviewed said they enlisted the assistance of intermediaries during their journeys. For example, Hassan, a Syrian refugee settled in Toronto, who was active in the 2015 protests in Syria, was informed by his contacts that his name appeared on one of the regime's apprehension lists. Learning of this news, he immediately began planning his journey to Lebanon. To negotiate the regulatory effects of inter-state mobility, Hassan journeyed through military checkpoints, hid from border snipers at the Syrian-Lebanese border, bypassed border security personnel and did so through the assistance of an intermediary, a 'smuggler'. In his words,

> I sold my belongings ... and I paid all the money to a smuggler, a taxi driver who can drive me safely to the Lebanese border. By paying this money [200,000 Syrian pounds, about US$400] to the personnel at the checkpoints on the road, they didn't check my ID, because if they did, they would put it on the computer and they would know that they want this person. So, I paid this money so that they don't check my ID at the Lebanese border. ... At the Lebanese border, I had to go out from the car and walk between the mountains for one and a half hours. I was protected by the Lebanese army because they got some money from the driver too. Before that, I hid myself really well from the snipers – the Syrian snipers who are on the border; they will shoot anyone who tries to flee. ... I hid myself, then met a soldier, a Lebanese soldier, who walked me near the border, because the driver wasn't sure if he would see his man at the border.

Notably, Hassan's journey through military checkpoints and the Syrian-Lebanese border involved strategies of communication, organization and border negotiation, which allow him and many others struggling to flee Syria to assert themselves not as victims, but as rights-bearing subjects, even if they are not authorized to do so (see Oliveri, 2015, on migrant struggles in Italy). In these journeys, the aid of an intermediary challenges the paternalistic vision of the relationship between the migrant and the smuggler, which denies migrants their agency to make journey plans and to give consent to the smuggling activity (see Isleyen, 2018, p. 13, on Turkish border crossings). Overall, such journey experiences highlight the politics and complexity of unauthorized mobility and the level of decision-making people on the move have not only during border crossings, but also under conditions in which the use of a human smuggler is the only hope they have to escape the social injustices of war. In this regard, struggles for migrant justice involve enlisting the help of smugglers, negotiating border crossings and willingly undertaking dangerous migratory journeys in an effort to access temporary safety and protection.

Syrian refugees I spoke with consider their journeys as challenging the grave injustices perpetrated by the Syrian state and its bordering practices, but living in host states presented them with other challenges that cast dark shadows over their future prospects. For example, Syrians' precarious status in Lebanon is shaped by the state's refusal to classify Syrians as 'refugees', using instead the term 'displaced persons', a category that does not offer legal protection for Syrians in a country that is not a signatory to the 1951 Refugee

Convention that relates to the Status of Refugees and its 1967 Protocol (Sanyal, 2017, pp. 117–118). Since 1963, the Lebanese state has permitted UNHCR to operate in the country and to take key responsibility for protecting and assisting the country's non-Palestinian refugee population (Janmyr & Mourad, 2018, p. 4). In this regard, one Syrian woman interviewee commented on the precarious conditions of her life in Lebanon:

> We [six of us] were crowded in two rooms, and my father was with me, he is 78. … We lived on UNHCR: they gave us $100 [US] for the house and 150,000 Lebanese Pounds [about US$100] for each person. The rent for the house would reach $326 [US], with other things like electricity.

Other Syrian interviewees who lived in Beirut expressed their concerns about the lack of legal protection for refugees and strict visa and work protocols, such as the requirements of a sponsor and payment of US$200 for every residency status renewal. These financial costs are particularly burdensome for Syrians, most of whom have stopped renewing their residency status and live in the country illegally. One man who worked in the construction and service industry and was unable to renew his residency status understood that he could be deported to Syria. In his words: "I was illegally in Lebanon. I didn't have residency. If I got caught they would arrest me or send me back".

During interviews, the majority of Syrians spoke about their precarious living conditions in Lebanon, Jordan and Turkey, including the lack of permanent residency and formal citizenship, inadequate access to employment, education and healthcare, obstacles to legal and social assistance (see Biehl, 2015; Ilcan, Rygiel & Baban, 2018), poor living and working conditions (see Landau, 2006; Pasquetti, 2015) and prevailing anti-refugee and anti-foreigner sentiments (see Şenoğuz, 2017, p. 166). Such living conditions produced increased uncertainty. For example, the Palestinian families who traveled from Syria to Jordan spoke about their extended time in Zaatari refugee camp, which houses over 100,000 refugees and is evolving into a permanent settlement. Given that Jordan hosted a large number of Palestinian refugees following the 1948 Arab-Israeli conflict, has a complex history of violence with Palestinians and is not a signatory to the 1951 Refugee Convention, Palestinian-Syrians in Jordan experience a decreased standard of living (see Sanyal, 2017, p. 120), including insufficient food and substandard accommodation as well as social injustices such as long-term confinement in cramped spaces and the denial of identity cards, resulting in the lack of access to basic social services. These indignities prompted camp protests against refugees' forced confinement in Zaatari and demands for fair treatment and access to necessary resources. As interviewees emphasized, the protestors rejected their invisibility and confinement by making their bodies visible and their claims noticeable in the spaces of Zaatari. Such struggles for migrant justice, no matter how small or infrequent, can play a crucial role in the manner in which refugees constitute themselves as political subjects and can serve as a basis for mobilizing solidarity and building alliances within and beyond camps.

The ways in which Syrians negotiate journeys out of Syria and struggle for migrant justice suggest they are not victims incapable of advocating for a politics of mobility justice. Instead, their struggles, like those of other mobile social groups (see Biehl, 2015; Dağtaş, 2017; Johnson, 2014), speak to the political agency of migrants and refugees in making decisions, negotiating bordering practices and mobilizing for social and political change. With many Syrian refugees facing unjust social and legal conditions in Lebanon, Jordan and Turkey, those I interviewed made the decision to continue their journey to Canada to access greater protection, safety and formal citizenship.

Conclusion

My analysis contributes to the scholarly work on mobility and justice (e.g., Cook & Butz, 2016; Cranstron, Schapendonk & Spaan, 2018; Ilcan, 2013a; Sheller, 2014) by focusing on Syrian refugees' encounters with unjust bordering practices during their migrant journeys (see Collyer, 2010; Innes, 2015; Mainwaring & Brigden, 2016), their challenges to those state mobilities (see Hess, 2017; Üstübici, 2016) and the political and ethical implications of those struggles and challenges, particularly the cultivation of demands for mobility justice. I have argued that migrant and refugee movements from Syria to nearby host states are not linear, but highly fluid political processes that involve migrant demands to move and leave, to negotiate bordering practices at military checkpoints and territorial borders and to mobilize their membership for inclusion or recognition in host states. Even in the face of state migration policies, massive conflicts and bordering and containment practices, migrants and refugees have engaged in what I have called struggles for migrant justice. Such struggles are imbricated in relations of power and demands for mobility justice that require the transformation of existing inequitable distributions of social, economic and political justice. In the migration and border context, mobility justice entails the development of a sustainable social, economic and political system of care, protection, rights and citizenship that will advantage everyone on the move, including vulnerable and poor migrant populations who have been forced to escape and relocate due to massive conflict, displacement and fear of persecution. In this regard, mobility justice includes and goes beyond the movements of state policies, migrants, refugees and asylum seekers, and human smuggling, to analyze diverse mobilities and immobilities, their complexities and transformations and their ties to issues of social justice that can produce new kinds of mobile situations, spaces and subjectivities.

In light of 'mobility turn' insights about the relationship among mobilities, power and (in)justice and the justice-related implications of the mechanisms through which mobilities are produced, controlled and contested, more ethnographic-based migration and border research is needed that frames migratory journey stories and migrant, community and border advocacy networks through the lens of mobility justice. A more comprehensive understanding of mobility justice in the migration context will push its conceptualization in new directions to create theoretical and ethical bridges between diverse actors and the political spaces of action and transformation.

Acknowledgements

The research for this paper was supported by a SSHRC Partnership Development Grant (2016–2019). I acknowledge with much appreciation the Syrian refugees in Ontario, Canada, who participated in this study and thank them for their time and valuable insights. I thank MA graduate students Zainab Abu Alrob and Nour Al Nasser for their research assistance. I thank Daniel O'Connor for his critical insights and the two editors, David Butz and Nancy Cook, for their very helpful suggestions and input on a previous version of this chapter.

References

Adey, P. (2016). 'Governing emergencies'. *Mobilities*, 11(1), 32–48.

Agier, M. (2011). *Managing the undesirables: Refugee camps and humanitarian government*. Cambridge: Polity Press.

Almustafa, M. (2018). 'Relived vulnerabilities of Palestianian refugees: Governing through exclusion', *Social and Legal Studies*, 27(2), 164–179.

Ataç, I., Heck, G., Hess, S., Kasli, Z., Ratfisch, P., Soykan, C. & Yilmaz, B. (2017). 'Contested b/orders: Turkey's changing migration regime'. *Movements: Journal for Critical Migration & Border Regime Studies*, 3(2), 9–22.

Bargu, B. (2017). 'The silent exception: Hunger striking and lip-sewing'. *Law, Culture & the Humanities*, 1–28. doi:10.1177/1743872117709684.

Biehl, K. S. (2015). 'Governing through uncertainty: Experiences of being a refugee in Turkey as a country for temporary asylum'. *Social Analysis*, 5(1), 57–75.

Brodie, J. (2013). 'Mobility regimes: The short life and times of North America's security and prosperity partnership'. In S. Ilcan (Ed.), *Mobilities, knowledge and social justice* (pp. 131–151). Montreal, QC: McGill-Queen's University Press.

Carlson, R. & Gallagher, M. (2015). 'Humanitarian protection for children fleeing gang-based violence in the Americas'. *Journal on Migration & Human Security*, 3(2), 129–158.

Collyer, M. (2010). 'Stranded migrants and the fragmented journey'. *Journal of Refugee Studies*, 23(3), 273–293.

Cook, N. & Butz, D. (2016). 'Mobility justice in the context of disaster'. *Mobilities*, 11 (3), 400–419.

Cranstron, S., Schapendonk, J. & Spaan, E. (2018). 'New directions in exploring the migration industries: Introduction to special issue'. *Journal of Ethnic & Migration Studies*, 44(4), 543–557.

Cresswell, T. (2006). *On the move*. London: Routledge.

Dağtaş, S. (2017). 'Whose Misafirs?: Negotiating difference along the Turkish–Syrian Border'. *International Journal of Middle East Studies*, 49(4), 661–679.

Doty, R. (2011). 'Bare life: Border-crossing deaths and spaces of moral alibi'. *Environment & Planning D: Society & Space*, 29(4), 599–612.

Genç, F., Heck, G. & Hess, S. (2018). 'The multilayered migration regime in Turkey: Contested regionalization, deceleration and legal precarization'. *Journal of Borderlands Studies.* doi:10.1080/08865655.2017.1344562.

Gerard, A. & Pickering, S. (2014). 'Gender securitization and transit: Refugee women and the journey to the EU'. *Journal of Refugee Studies,* 27(3), 338–359.

Hess, S. (2017). 'Border crossing as act of resistance: The autonomy of migration as theoretical intervention into border studies'. In M. Buler, P. Mecheril & L. Brenningmeyer (Eds.), *Resistance: Subjects, representations, contexts* (pp. 87–100). New York: Columbia University Press.

Ilcan, S. (2017, November). 'Migrant journeys, border zone resistance and affective encounters: Ethnographic recollections of Syrian refugee movements'. Paper presented at the Bordering Practices in Migration & Refugee Protection conference, Waterloo, ON.

Ilcan, S. (2013a). *Mobilities, knowledge and social justice.* Montreal, QC: McGill-Queen's University Press.

Ilcan, S. (2013b). 'Introduction: Mobilities, knowledge and social justice'. In S. Ilcan (Ed.), *Mobilities, knowledge and social justice* (pp. 3–24). Montreal, QC: McGill-Queen's University Press.

Ilcan, S. (2013c). 'Paradoxes of humanitarian aid: Mobile populations, biopolitical knowledge and acts of social justice in Osire refugee camp'. In S. Ilcan (Ed.), *Mobilities, knowledge and social justice* (pp. 177–206). Montreal, QC: McGill-Queen's University Press.

Ilcan, S., Rygiel, K. & Baban, F. (2018). 'The ambiguous architecture of precarity: Temporary protection, everyday living and migrant journeys of Syrian refugees'. *International Journal of Migration & Borders,* 4(1–2), 51–70.

Innes, A. J. (2015). 'The never-ending journey? Exclusive jurisdictions and migrant mobility in Europe'. *Journal of Contemporary European Studies,* 23(4), 500–513.

Isleyen, B. (2018). 'Turkey's governance of irregular migration at European Union borders: Emerging geographies of care and control'. *Environment & Planning D: Society & Space.* doi:10.1177/0263775818762132.

Janmyr, M. & Mourad, L. (2018). 'Modes of ordering: Labeling, classification and categorization in Lebanon's refugee response'. *Journal of Refugee Studies.* doi:10.1093/fex042.

Johnson, H. (2014). *Borders, asylum and global non-citizenship: The other side of the fence.* Cambridge: Cambridge University Press.

Lambert, H. (2017). 'Temporary refuge from war: Customary international law and the Syrian Conflict'. *International & Comparative Law Quarterly,* 66(3), 723–745.

Landau, L. (2006). 'Protection and dignity in Johannesburg: Shortcomings of South Africa's urban refugee policy'. *Journal of Refugee Studies,* 19(3), 308–327.

Leenders, R. & Heydemann, S. (2012). 'Popular mobilization in Syria: Opportunity and threat, and the social networks of the early risers'. *Mediterranean Politics,* 17(2), 139–159.

Little, A. (2015). 'The complex temporality of borders: Contingency and normativity'. *European Journal of Political Theory,* 14(4), 429–447.

Mainwaring, C. & Brigden, N. (2016). 'Beyond the border: Clandestine migration journeys'. *Geopolitics,* 21(2), 243–262.

Mountz, A. (2011). 'Specters at the port of entry: Understanding state mobilities through an ontology of exclusion'. *Mobilities,* 6(3), 317–334.

Nevins, J. (2002). *Operation gatekeeper: The rise of the 'illegal alien' and the making of the U.S.–Mexico boundary.* New York: Routledge.

Oliveri, F. (2015). 'Subverting neoliberal citizenship: Migrant struggles for the right to stay in contemporary Italy'. *ACME: An International E-Journal for Critical Geographies*, 14(2), 492–503.

Pallister-Wilkins, P. (2015). 'The humanitarian politics of European border policing: Frontex and border police in Evros'. *International Political Sociology*, 9(1), 53–69.

Pasquetti, S. (2015). 'Negotiating control'. *City*, 19(5), 702–713.

Pickering, S. & Weber, L. (2006). *Borders, mobilities and technologies of control*. Dordrecht: Springer.

Rygiel, K. (2013). 'Mobile citizens, risky subjects: Security knowledge at the border'. In S. Ilcan (Ed.), *Mobilities, knowledge and social justice* (pp. 131–151). Montreal, QC: McGill-Queen's University Press.

Sanyal, R. (2017). 'A no-camp policy: Interrogating informal settlements in Lebanon'. *Geoforum*, 84, 117–125.

Schapendonk, J. & Steel, G. (2014). 'Following migrant trajectories: The im/mobility of sub-Saharan Africans en route to the European Union'. *Annals of the Association of American Geographers*, 104(2), 262–270.

Şenoğuz, H. (2017). 'Border contestations, Syrian refugees and violence in the southeastern margins of Turkey'. *Movements: Journal for Critical Migration & Border Regime Studies*, 3(2), 163–176.

Sheller, M. (2018). *Mobility justice: The politics of movement in an age of extremes*. New York: Verso.

Sheller, M. (2017). 'From spatial turn to mobilities turn'. *Current Sociology*, 65(4), 623–639.

Sheller, M. (2014). 'The new mobilities paradigm for a live sociology'. *Current Sociology*, 62(6), 789–811.

Squire, V. (2014). 'Desert "trash": Poshumanism, border struggles and humanitarian politics'. *Political Geography*, 39, 11–21.

Stasiulis, D. (2013). 'Contending frames of security and citizenship: Lebanese dual nationals during the 2006 Lebanon war'. In S. Ilcan (Ed.), *Mobilities, knowledge and social justice* (pp. 25–58). Montreal: McGill-Queen's University Press.

Thorleifesson, C. (2016). 'The limits of hospitality: Coping strategies among displaced Syrians in Lebanon'. *Third World Quarterly*, 37(6), 1071–1081.

Turner, L. (2015). 'Explaining the (non-)encampment of Syrian refugees: Security, class and the labour market in Lebanon and Jordan'. *Mediterranean Politics*, 20(3), 386–404.

Tyler, I. & Marciniak, K. (2013). 'Immigrant protest: An introduction'. *Citizenship Studies*, 17(2), 143–156.

Urry, J. (2007). *Mobilities*. Cambridge: Polity Press.

Üstübici, A. (2016). 'Political activism between journey and settlement: Irregular migrant mobilization in Morocco'. *Geopolitics*, 21(2), 303–324.

Walters, W. (2006). 'Border/control'. *European Journal of Social Theory*, 9(2), 187–203.

Weber, L. & Pickering, S. (2011). *Transnational crime, crime control and security*. Basingstoke: Palgrave Macmillan.

Williams, J. (2015). 'From humanitarian exceptionalism to contingent care: Care and enforcement at the humanitarian border'. *Political Geography*, 47, 11–20.

5 Transportation exploitation, mobility and social justice

A critical analysis

Gerard C. Wellman

Introduction

Mobility is critical to a functioning society as it encompasses "both the large-scale movements of people, objects, capital and information across the world, as well as the more local processes of daily transportation, movement through public space and the travel of material things within everyday life" (Hannam, Sheller & Urry, 2006, p. 1). I focus on "local processes of daily transportation" to argue that without mobility provision through transportation, individuals have difficulty accessing basic services, engaging in cultural activities, involving themselves in employment and educational opportunities and pursuing actions that enrich and promote life chances (Hanson, 1995; Sheller, 2014). Mobility provision reflects transportation planners' best attempts to determine how citizens share space and time (Martens, 2012) and embodies an unshakably utopian idea that through effective and efficient planning, we can mold the world to augment societal wellbeing (Friedmann, 1987; Healey, 2003, 2015). Few areas of public policymaking believe as strongly in the possibility of a planned utopia, accompanied by positive externalities, in the lives of ordinary people (Zimmerman, 2012).

Providing mobility requires planning and policymaking that ensures movement in safe modes of travel that minimize costs to others (Steinemann, Apgar & Brown, 2005) and investments in transportation infrastructure – roads, sidewalks, rail lines – that make personal mobility possible. In other words, mobility provision is a policy problem requiring public will, group decision-making and the allocation of scarce public resources to determine how individuals share space and time in movement and how transportation costs and benefits are fairly distributed among populations (Sheller & Urry, 2000). Because space, time and public budgets are finite, and transportation infrastructure has benefits (e.g., a road provides mobility options) and costs (e.g., reduced air quality for residents living adjacent to the road) that should be fairly distributed, mobility is a public and political problem, with social justice implications.

Of further concern are the power relations exercised through transportation planning and mobility provision (Portugali & Alfasi, 2008). Few decisions are as influential as those made by transportation administrators and policy-makers, who control how people get to work and whether a particular transit line will be expanded or cut. They also curtail or create access to educational, cultural, healthcare and employment opportunities for differently gendered, racialized and classed groups (Preston & Raje, 2007). Indeed, they determine whether entire populations are able to leave home, and how safely; raising speed limits or refusing to install pedestrian crosswalks have life-changing consequences. These exercises of power outside an idealized democratic con-text are often unchecked, even unnoticed, yet have produced transportation exploitation – the use of policy to unevenly distribute mobility options that physically contain marginalized populations, constraining their access to social resources and producing social exclusion. This exploitation denies those groups the "personal freedom of mobility" (Sheller, 2008, p. 30), the ability to determine how, where and when they can physically move. This mobility-related injustice has implications for individuals' life chances, agency and social inclusion. While *de jure* transport segregation is rare, planning and policymaking deference to the automobile results in *de facto* segregation between those who do and do not own vehicles, resulting in poor and mar-ginalized communities' differential access to social opportunities. Among its many goals, mobility justice seeks to eliminate this transportation exploitation.

I draw on theories of social justice to examine how transportation provision often reduces access to mobility-related goods, thereby serving as a tool of exploitation that preserves existing class structures and promulgates negative mobility outcomes for underserved communities. I discuss these conceptual tools before analyzing transportation exploitation, the effects of unjust allo-cations of transportation resources and perspectives of transportation admin-istrators who are seeking to provide just outcomes to the communities they serve.

Social justice and mobility

Transportation policy in the Global North, which privileged the automobile throughout the 20th century, constrains mobility for some segments of society while enhancing it for others (Martens, 2006). Accordingly, the communities least likely to have broad daily transportation options disproportionately bear the costs of others' mobility in a mismatch of costs and benefits (Bullard, 2003). These mobility-related disparities are linked to the question of social inequality and perspectives on social and environmental justice.

The relationship between mobility and social justice is variable because the concept of social justice is multidimensional. However, it is generally under-stood as the equitable distribution of scarce societal resources within a society (Barr, 1998, p. 452). Political orientations affect the term's specific applica-tions in certain situations, giving it an aura of ambiguity and rendering a specific definition useless (Piachaud, 2008, p. 33; Wolff, 2008, p. 17). Some authors call it a "feel good" term with little practical resonance (Hayek, 1976, p. 66; Piachaud, 2008, p. 33); others like Minogue (1998), Barry (2005) and Rawls (1971) argue that social justice is a universal concept with concrete principles applicable to all societies; still others like Miller (1976) and Walzer (1983) insist it is interpretivistic in nature and subject to the dominant culture of different societies.

Discussing the British welfare state, Barr (1998) argues that a just society assists citizens in choosing the social arrangements they desire, like the size and extent of the welfare state, the scope and desirability of military interventions or – in a flagrant extension of his writing – the accessibility of daily transport mobility and the public policy mechanisms employed to create it. This 'application' of justice principles speaks to claims made by Rawls (2001), Harvey (2009), Piachaud (2008) and Walzer (1983) that understanding justice merely in a theoretical way is insufficient, because justice has concrete, practical implications for policy formulation, urban development decisions and lived experience. As Barry (2005, p. 4) argues, "[w]e need the right theory of social justice if we are to get the right answers"; employing an impractical concept of justice may have drastic consequences for the life chances of the poorest and least mobile members of society.

Although there is no agreement on how to achieve this goal, some approaches to social justice are particularly relevant to achieving equitable daily mobility. Rawls (2001) and other liberals who advocate for mixed-market approaches to social problems would favor the use of free market principles in transportation planning to enhance daily mobility for the poorest segments of society, meeting the transportation needs of the least well-off first. Collectivists, similarly, would favor government intervention to provide mobility, sometimes including free public transit, to ensure the broadest possible distribution of economic equality and mobility. However, Walzer (1983), who is likewise concerned with the distribution of social 'goods', argues that public subsidies (re)produce class inequalities when some groups acquire more benefits and others more costs (Martens, 2012). According to Bullard, Johnson and Torres (2000, p. 46),

> transportation is a key component in addressing poverty, unemployment and equal opportunity goals, and ensuring equal access to education, employment and other public services. In the real world, transportation decisions do not have the same impact on all groups. Costs and benefits associated with transportation developments are not randomly distributed.

This uneven distribution of access to and the costs and benefits of daily transportation can become a source of social conflict; some groups monopolize transportation access and convert it into other goods, ensuring access to education, jobs, healthcare and even time itself (Walzer, 1983). For instance, Domosh and Seager (2001) argue that men disproportionately access daily transportation, thereby entrenching gender inequalities. Bullard, Johnson and Torres (2000) suggest that white privilege structures transport access, with implications for racial marginalization. In tandem with these analyses, I argue that access to daily transportation is also a *classed* phenomenon in which privileged classes use transportation systems to improve their life chances at the expense of poor communities, without regard for social justice.

Transportation exploitation

Transportation policy has historically appeared more like a tool of societal conquest than a planned progression toward utopia. Worldwide, it has rightly been characterized as a tool of sexism, racism, ageism and ableism through the withholding of mobility from marginalized social groups. While mobility globally has rarely been withheld to produce *de jure* spatial segregation, private automobiles created new spatial and temporal inequalities in Western societies (Bullard, Johnson & Torres, 2000; Domosh & Seager, 2001; Sachs, 1992). Domosh and Seager (2001, p. 114) argue that "[w]ealthier people move through space more easily than their poorer counterparts: simply put, wealthy people fly, poor people take the bus". In other words, wealth creates a range of transportation options from which to choose, and some options, like reliable automobiles, increase a person's control over time and space. Accordingly, Domosh and Seager (2001, p. 114) note that people waiting for the bus that never arrives are usually not from the privileged classes.

Control over travel method and time is a frequent theme in transportation literature (e.g., Ureta, 2008). Research shows that wealthier individuals, who more easily control how and how fast they reach their destinations, usually favor policy that shifts funds away from public transit and toward investments that make driving automobiles more convenient (Sachs, 1992). Their disproportionate political and transport policymaking influence is manifest in myriad ways that secure resources for their preferred modes of travel (Kenyon, Rafferty & Lyons, 2002; Ureta, 2008). The resulting negative effects for people who do not own vehicles are evidence of exploitation through transportation planning.

Erik Olin Wright (1997, p. 10), in his book *Class Counts*, defines exploitation on the basis of three criteria: first, "the material welfare of one group of people causally depends on the material deprivations of another"; second, "the causal relation [in the first criterion] involves the asymmetrical exclusion of the exploited from access to certain productive resources"; and, finally, "the causal mechanism which translates exclusion [in the second criterion] into differential welfare [in the first criterion] involves the appropriation of the fruits of labor of the exploited by those who control the relevant productive resources". I translate this neo-Marxist economic understanding of class exploitation into a critique of daily transportation policy in the following way: first, the mobility of one social group causally (and antagonistically) depends on restricting the mobility of others (e.g., the addition of a turning lane that makes driving more convenient lengthens the section pedestrians must cross at an intersection, exposing them to more oncoming traffic); second, this causal relationship prohibits exploited groups from using other means of mobility based on their social position to access social resources (e.g., the purchase and upkeep of private automobiles are cost-prohibitive for poor communities); and, third, mobility exclusions engendered through transportation policy differentially benefit economically powerful groups, reproducing their privilege (a similar process occurs in land use decisions that require people to purchase automobiles before they can participate in the workforce and be economically productive) (Logan & Molotch, 1987). The act of severing individuals from cultural, employment, healthcare and education resources because they are too poor to own automobiles – and of making their walking and public transit journeys precarious and lengthy due to massive public investments in driving infrastructure – neatly demonstrates transportation exploitation (for a historical perspective, see Roth, 2004).

Lutz and Fernandez (2010, p. 8) illustrate Wright's third point; as of 2010, six of the largest corporations in the United States were automobile or oil-related businesses. Corporate elites, therefore, have vested interests in ensuring that policymakers dedicate maximum funding to road expansion, rather than public transportation systems that provide greater social justice benefits. As roads stretch well beyond cities, property-owning elites exert influence to ensure those roads increase accessibility to, and thus the value of, their developable land. By sinking public money into infrastructure that requires a significant private investment to use (i.e., an automobile), the poor are exploited and excluded from transportation policymaking and policy benefits, with drastically negative consequences:

> The car system redistributes wealth upward, magnifying inequality. ... First, people in poor, carless households often face extreme difficulty in getting and keeping employment because it is challenging for them simply to get from their homes to available jobs. Second, working- and middle-class people are being held back from economic advancement or pushed further into debt through car ownership. Cars are an expensive and depreciating asset for which there remains pervasive discrimination in pricing and financing. What these first two factors mean is that the poor can't live without the car and they can't live with it. Third, at the same time, the car consolidates and elevates the status of the wealthy. Some of the very wealthiest families in the nation are or have been the beneficiaries of the auto-industry, or of windfall gasoline and banking profits reaped from the car-and gasoline-buying public.
>
> (Lutz & Fernandez, 2010, p. 102)

Transportation policy, particularly the debate about whether or not to invest in public transit, entails social control by elites through exploitation that limits the mobility of marginalized underclasses, constraining their associations and access to resources and engendering social exclusion (Domosh & Seager, 2001, p. 115). This process, in effect, restricts *social* mobility by limiting *physical* mobility – redistributing power toward the powerful – which is not a new phenomenon.

Withholding mobility

One of the earliest recorded forms of mobility restriction as a proxy for social restriction is the use of footbinding in 10th-century China, which crippled women into dependent domesticity and servitude (Bossen, Xurui, Brown & Gates, 2011). While Saudi Arabia's prohibition (until 2017) of women drivers was notorious, other jurisdictions, including Gaza, still constrain women's driving without a male 'chaperone' (Hadid & Waheidi, 2016; Harris, 2016). As Domosh and Seager (2001, p. 115) point out,

> The control of women's [mobility] has long preoccupied governments, families, households and individual men. It is hard to maintain patriarchal control over women if they have unfettered freedom of movement through space. Sometimes this control is exercised bluntly and brutally. ... Tight corseting, high heels, hobble skirts, the veil, prohibitions against women riding bicycles or horses, restrictions (legal or social) on women driving cars – all suggest the extent to which "keeping women in their place" is often a literal undertaking.

As motorization provided new modes of mobility, new restrictions were enacted to delineate which groups could use the technology, perpetuating mobility injustice and social exclusion. For instance, in 1892, Homer Plessy, an African-American customer on the East Louisiana railway, purposefully rode in a railcar exclusively reserved for white patrons (Bullard, 2003). He was arrested, and his case was eventually adjudicated before the Supreme Court, which used the opportunity to codify the 'separate but equal' doctrine that remained in place for 60 years. Louisiana's insistence that African-American passengers sit in different railcars was, in the view of the court, perfectly legal and permissible, opening up avenues for segregationists to withhold mobility through Jim Crow laws related to transportation modes (Martin, 1998).

But resistance to this immobilization was widespread. For instance, Rosa Parks, an African-American woman from Montgomery, Alabama, challenged racial segregation in public spaces through civil disobedience on December 1, 1955. Defying the city's rule that African-American riders must sit in the back of transit buses, Ms. Parks' arrest sparked a boycott of the city's bus system and cemented transport-related civil disobedience as useful in the quest for social justice (Banks, 1994; Parks & Haskins, 1992). Moreover, in 1961, the Freedom Riders used transit buses to protest segregation in public transit. Greeted by violent Southern segregationists, the Freedom Riders demonstrated how keeping racially marginalized populations 'in their place' could be cruelly, effectively and efficiently achieved through transportation policy. And the Watts riot in Los Angeles, California, in 1965 shone a spotlight on the dismal conditions on public transit systems in the United States. The subsequent McCone Commission, created by the Governor of California to identify and investigate the cause of the riot, reported that the

> inadequate and costly public transportation currently existing throughout the Los Angeles area seriously restricts the residents of the disadvantaged areas such as south central Los Angeles. This lack of adequate transportation handicaps them in seeking and holding jobs, attending schools, shopping and in fulfilling other needs. It has had a major influence in creating a sense of isolation.
>
> (The Governor's Commission on the Los Angeles Riots, 1965, p. 65)

The Commission wrote that the nearest county hospital was a two-hour trip by bus; reaching the Youth Employment Training Center took over 90 minutes (Sheldon & Brandwein, 1973, p. 7). Because of the US government's policy priorities at the time, and the result of the riots and the McCone Commission findings, public transportation received greater subsidies to address transit inequities in the inner city – subsidies that continue today (Deka, 2004). Yet the subsidies have arguably not achieved transportation equity, and issues of social justice in transportation continue to be outweighed by policy provision for private transport.

Bullard (2003) contends that isolating the poor through reduced physical mobility is exploitative and compounds the problems of distressed communities. Garrett and Taylor (1999, p. 8) discuss a former Cleveland planning director who saw the urban development process in Ohio as inherently oppressive of the economically disadvantaged. Poor communities, he argued, were isolated through underfunded, poorly performing public transit, while there was significant public investment in urban highways that facilitated white flight and a hollowing out of central cities (Garrett & Taylor, 1999, p. 8). Exploitation of poor communities is evidenced in their neighborhood isolation and ineffective public transit service and in the need for automobiles to meet daily mobility requirements. Further, disinvestment in urban public transit for massive highway subsidies effectively subsidized the middle-class flight to the suburbs, at the literal expense of the urban poor. Between 1947 and 1970 in the United States, for example, federal spending on transportation varied significantly by mode: while $795 million was spent on public transportation, $12 billion was allocated for airport development, $6 billion for ports and water-based transportation and $58 billion for infrastructure for cars, including roads and highways (Davies, 1975, p. 36). In other words, during the civil rights era in the United States, the federal government spent $73 on roads for every $1 it spent on urban public transit.

This trend continues. Presently in the United States, 80% of all government spending on transportation is dedicated to car-based travel; the remainder is channeled to public and active transportation (Bullard, 2003; Sanchez, Stolz & Ma, 2003, p. 14). Transportation also occupies a significant portion of household spending in the United States; this accounted for 10% of household funds in the United States in 1960; in the early 2000s, 17.1% of household funds in New England and 20.8% in the South were used for this purpose (Bullard, 2003; Litman, 2007). Such spending is regressive, with the poorest quintile of US residents committing 40% of their take-home pay solely to mobility expenses (Bullard, 2003; Litman, 2007). Many households are unable to afford a private vehicle – 17% of Latino and Asian households and 24% of African-American households do not own one (Sanchez, Stolz & Ma, 2003, p. 9; US Department of Transportation, 2004). Among all households whose net income is less than $20,000 a year, 27% do not own an automobile (US Department of Transportation, 2004). Among those over the age of 65, 21% do not drive, with implications for their healthcare and social activities; 15% restrict trips for healthcare services, and 65% restrict trips for family and social engagements (Bailey, 2004, p. 3). Thus, "lack of car ownership and inadequate public transit service in many central cities and metropolitan regions with high proportions of 'captive' transit dependents exacerbate social, economic and racial isolation, especially for low-income people of color" (Bullard, 2003, para. 6).

Davies (1975, p. 32) notes that publicly funded urban freeways that connect urban downtowns with low-density suburbs were usually built through neighborhoods chosen by planners because those "residents had less political influence, and were less likely to organize an effective protest". In Nashville, Tennessee, for example, Interstate 40 sliced through a poor African-American community, "leaving more than fifty dead end streets, forever dividing what had been a cohesive neighborhood" (Davies, 1975, p. 32). The "construction of highways became one of the largest civil engineering feats of the century, but also tore apart the existing fabric of the United States" (Reed & Sen, 2005, p. 39), leaving urban minority communities more isolated and separated than before their construction. Such neighborhoods have since been prone to the negative health effects of living immediately adjacent to wide, busy freeways, reinforcing compounding layers of negative externalities (Newman & Kenworthy, 2015). Ironically, communities with the greatest access to urban freeways are usually those with the least access to personal automobiles needed to traverse such roads.

Disinvesting in the mobility needs of the poor consists of reducing transit services and daily transportation infrastructure like sidewalks and crosswalks (Lowe, 2016; Kravetz & Noland, 2012). In my research on transportation administrators' perspectives on social justice, one administrator working at a mid-sized transit agency in upstate New York described a nearby suburban county that likes to pretend they don't have any poor people, and they hide them in these trailer parks off ... a main road that you can't serve by transit in any kind of reasonable way. ... [This county] has the worst land use policies in terms of sprawl and a lack of transit-supportive land use policies. ... No sidewalks [for those who do not own cars].

How, she rhetorically asked, can the transit agency serve trailer park residents if the agency is physically restricted from accessing them and the residents are physically restricted from walking to transit stops because there are no sidewalks?

Implicit in such mobility restrictions is the idea that automobiles constitute the 'correct' way to travel. Cultural messages reinforce this idea; outside large cities, public transportation is seen as relevant only to those populations too poor to afford cars. As Angrosino's (1994, p. 21) ethnographic research on urban poverty demonstrates, "the bus is the very embodiment of stigma, the slow inconvenient transport of the poor, the powerless and the socially marginal". One Louisiana administrator from a mid-sized transit agency told me there are "perceptions of stigmas of people ... [that] just because you ride the bus you're a low-life". Another from New Mexico claimed during an interview that "there is a stereotype ... about the population of bus riders, that they are, you know, socially and economically deprived". Choice riders – those with transportation alternatives – "were afraid of transit. They were afraid of the people that were bus riders". To eradicate the stigma of bus ridership, this agency took "great steps to diversify the ridership" by adding "Wi-Fi on ... [express] routes to try to draw more of a business community". If privileged classes utilize transit, according to this administrator's logic, then the stigma will be erased. Social and mobility justice is partially achieved, then, when "somebody can go into a room and say, 'well I'm a bus rider' and not be looked at in a negative way".

Another administrator claimed that bus ridership stigma is an inevitable outcome of reduced funding, which leads to greater social stigma in a dysfunctional cycle:

When you try to operate a system based on such a shoestring [budget], you put the services more to where they're perceived to be needed. ... You try to meet those needs first. It's kind of a logical way to operate. But [what happens is] you create a perception problem that the system is only for the dispossessed. When you do that, it just completely destroys the system, and you continue to shrink and shrink and shrink.

An administrator from Ohio saw the stigma of bus ridership as pervasive and indicative of racism, echoing Wright's (1997) and Roth's (2004) perspectives on exploitation:

We still have some lingering perceptions here that the buses are just for poor black people, and these are people who are unemployed and can be dangerous, and I'm just telling you flat that this is the way people think to a large degree, especially in the suburbs and sometimes even in the business community. And in fact [they] even blame the demise of a shopping center on the fact that there was a bus stop there.

The provision of mobility and retail access to economically disadvantaged communities (at least, the communities most likely to be stigmatized as such) led to accusations that the transit agency and the daily transportation access it provided were actually *frustrating* economic development.

These illustrations reveal the extent to which the daily transportation needs of poor communities are not only ignored, but also actively repressed. Walzer's notion of social justice as the equitable distribution of and access to goods like daily transportation demands that public policy decisions should not (re)produce differential mobilities on the basis of wealth. Yet we see that social *injustice* and Wright's criteria for exploitation have found new life in daily transportation policy.

Conclusion

Transportation policy in the United States has been neatly hijacked by groups seeking to maintain class and mobility hierarchies. Privileged populations that wield physical isolation as a tool for preserving existing class systems exploit vulnerable populations, producing considerable challenges for adopting socially just policies. This phenomenon is not new; historically, keeping someone 'in their place' socially has involved mobility restrictions.

My primary focus has been to demonstrate how the power relations exercised through transportation planning and mobility provision have produced such exploitation via the use of policy to unevenly distribute mobility options that physically contain marginalized populations, constraining their access to social resources and producing social exclusion. This exploitation denies these groups the freedom to determine how, where and when they move, constituting a mobility injustice. Are we 'free' if physical access to and distribution of transportation goods are denied to us? Does a society truly value freedom if, as a result of transportation planning and public funding decisions, individuals are trapped in their communities or homes, unable to access basic social resources? Drawing on the work of Walzer, Wright, Sheller and others for support, I strongly suggest not.

Transportation is crucial to a functioning society, yet the field in most Western societies has strayed by limiting its purview to one mode of transportation (private automobiles), thereby sidelining public, active and alternative modes that are more accessible across populations. The principles of mobility justice suggest that the field of transportation policy and planning needs a broader view of the mobility needs of a range of groups, especially the socially marginalized. This reorientation requires a return to the field's utopian roots and necessitates a broad and thorny conversation about what it means to share space and time, and whether or not basic mobility is a human freedom and right.

Advocating for mobility justice also necessitates a more fulsome understanding and evaluation of the everyday lived social consequences of *unjust* transportation planning for the people served by planners and policymakers. One administrator I interviewed arrived at a similar conclusion:

> Shortly after I came to the [agency] ... the revenues were down because the jobs were down. ... So we had to make a lot of service cuts. So we decided that we wanted to provide as much work trip service as possible. So we said "Okay, let's look at weekends, Sunday service. A lot of Sunday service is primarily used by people going to church. Well, it's better to cut that than it is to cut a job route".
>
> I can remember being in a public hearing, and we were talking about all these Sunday services that we were going to cut. At the end of the hearing, this elderly woman came up to me, and I was the hearing officer so to speak, the one making the presentation. This lady comes up to me, and I never will forget it. Little old lady, she's got a cane. She said, "Mister ... are you a Christian man?" I said, "Well, I like to think I am". She said, "Well, I hope you are because you'd better get down on your knees and pray to God that you don't die in your sleep and go straight to hell" and walked off.
>
> Putting that into perspective, I cut her Sunday route. That was her life. I mean her whole social network was around going to church on Sunday. Now she can't do that. Is that a social justice issue? Absolutely.

References

Angrosino, M. V. (1994). 'On the bus with Vonnie Lee: Explorations in life history and metaphor'. *Journal of Contemporary Ethnography*, 23(1), 14–28.

Bailey, L. (2004). 'Aging Americans: Stranded without options'. *American Public Transportation Association*. Retrieved from www.apta.com/resources/reportsandp ublications/Documents/aging_stranded.pdf.

Banks, J. A. (1994). 'Transforming the mainstream curriculum'. *Educating for Diversity*, 51(8), 4–8.

Barr, N. (1998). *The economics of the welfare state* (3rd ed.). Stanford, CA: Stanford University Press.

Barry, B. (2005). *Why social justice matters*. Cambridge: Polity Press.

Bossen, L., Xurui, W., Brown, M. & Gates, H. (2011). 'Feet and fabrication: Footbinding and early twentieth-century rural women's labor in Shaanxi'. *Modern China*, 37(4), 347–383.

Bullard, R. D. (2003). 'New routes to transportation equity: Why race still matters'. *Transportation Equity: A Newsletter of the Environmental Justice Resource Center at Clark Atlanta University*, 6(1). Retrieved from www.ejrc.cau.edu/transequ newsvol6.htm.

Bullard, R. D., Johnson, G. S. & Torres, A. O. (2000). 'Dismantling transportation apartheid: The quest for equity'. In R. Bullard, G. Johnson & A. Torres (Eds.), *Sprawl city: Race, politics and planning in Atlanta* (pp. 39–68). Washington, DC: Island Press.

Davies, R. O. (1975). *The age of asphalt: The automobile, the freeway and the condition of metropolitan America*. Philadelphia, PA: J. B. Lippincott Company.

Deka, D. (2004). 'Social and environmental justice issues in urban transportation'. In S. Hanson & G. Giuliano (Eds.), *The geography of urban transportation* (3rd ed.) (pp. 332–355). New York: The Guilford Press.

Domosh, M. & Seager, J. (2001). *Putting women in place: Feminist geographies make sense of the world*. New York: The Guilford Press.

Friedmann, J. (1987). *Planning in the public domain: From knowledge to action*. Princeton, NJ: Princeton University Press.

Garrett, M. & Taylor, B. (1999). 'Reconsidering social equity in public transit'. *Berkeley Planning Journal*, 13(1), 6–27.

Hadid, D. & Waheidi, M. (2016, February 23). 'In Gaza, bicycles are a battleground for women who dare to ride'. *The New York Times*. Retrieved from www.nytimes.com/2016/02/23/world/middleeast/gaza-women-on-bikes-face-a-long-road-to-acceptance.html.

Hannam, K., Sheller, M. & Urry, J. (2006). 'Mobilities, immobilities and moorings'. *Mobilities*, 1(1), 1–22.

Hanson, S. (1995). 'Getting there: Urban transportation in context'. In S. Hanson (Ed.), *The geography of urban transportation* (pp. 3–25). New York: Guilford Press.

Harris, E. (2016, June 1). 'Gaza women learning to drive may have a chaperone'. *National Public Radio*. Retrieved from www.npr.org/sections/parallels/2016/06/01/478361330/hamas-gaza-women-learning-to-drive-must-have-a-chaperone.

Harvey, D. (2009). *Social justice and the city*. Athens, GA: The University of Georgia Press.

Hayek, F. (1976). *Law, legislation and liberty* (2nd ed.). New York: Routledge.

Healey, P. (2015). 'Planning theory: The good city and its governance'. *International Encyclopedia of Social & Behavioral Science*, 18, 202–207.

Healey, P. (2003). 'Collaborative planning perspective'. *Planning Theory*, 2(2), 101–123.

Kenyon, S., Rafferty, J. & Lyons, G. (2002). 'Transport and social exclusion: Investigating the possibility of promoting inclusion through virtual mobility'. *Journal of Transport Geography*, 10(3), 207–219.

Kravetz, D. & Noland, R. (2012). 'Spatial analysis of income disparities in pedestrian safety in Northern New Jersey: Is there an environmental justice issue?' *Transportation Research Record: Journal of the Transportation Research Board*, 2320, 10–17.

Litman, T. (2007). 'Evaluating transportation equity'. *World Transport Policy & Practice*, 8(2), 50–65.

Logan, J. R. & Molotch, H. L. (1987). *Urban fortunes: The political economy of place*. Los Angeles: University of California Press.

Lowe, K. (2016). 'Environmental justice and pedestrianism: Sidewalk continuity, race and poverty in New Orleans, Louisiana'. *Transportation Research Record: Journal of the Transportation Research Board*, 2598, 119–123.

Lutz, C. & Fernandez, A. L. (2010). *Carjacked: The culture of the automobile and its effect on our lives*. New York: Palgrave Macmillan.

Martens, K. (2012). 'Justice in transport as justice in access: Applying Walzer's "spheres of justice" to the transport section'. *Transportation*, 39(6), 1035–1053.

Martens, K. (2006). 'Basing transportation planning on principles of social justice'. *Berkeley Planning Journal*, 19(1), 1–17.

Martin, W. E. (1998). *Brown v. Board of Education: A brief history with documents.* Boston, MA: Bedford/St. Martin's.

Miller, D. (1976). *Social justice.* Oxford: Oxford University Press.

Minogue, K. (1998). 'Social justice in theory and practice'. In D. Boucher & P. Kelly (Eds.), *Social justice: From Hume to Walzer* (pp. 253–266). New York: Routledge.

Newman, P. & Kenworthy, J. (2015). *The end of automobile dependence: How cities are moving beyond car-based planning.* Washington, DC: Island Press.

Parks, R. & Haskins, J. (1992). *Rosa Parks: My story.* New York: Puffin Books.

Piachaud, D. (2008). 'Social justice and public policy: A social policy perspective'. In G. Craig, T. Burchardt & D. Gordon (Eds.), *Social justice and public policy: Seeking fairness in diverse societies* (pp. 33–52). Bristol, UK: The Policy Press.

Portugali, J. & Alfasi, N. (2008). 'An approach to planning discourse analysis'. *Urban Studies*, 45(2), 251–272.

Preston, J. & Raje, F. (2007). 'Accessibility, mobility and transport-related social exclusion'. *Journal of Transport Geography*, 15(3), 151–160.

Rawls, J. (2001). *Justice as fairness: A restatement.* Cambridge, MA: Harvard University Press.

Rawls, J. (1971). *A theory of justice.* Cambridge, MA: Harvard University Press.

Reed, R. & Sen, S. (2005). 'Racial differences and pedestrian safety: Some evidence from Maryland and implications for policy'. *Journal of Public Transportation*, 8(2), 37–61.

Roth, M. (2004). 'Whittier Boulevard, Sixth Street Bridge and the origins of transportation exploitation in East Los Angeles'. *Journal of Urban History*, 30(5), 729–748.

Sachs, W. (1992). *For love of the automobile.* Berkeley, CA: University of California.

Sanchez, T. W., Stolz, R. & Ma, J. S. (2003). *Moving to equity: Addressing inequitable effects of transportation policies on minorities.* Cambridge, MA: The Civil Rights Project at Harvard University.

Sheldon, N. W. & Brandwein, R. (1973). *The economic and social impact of investments in public transit.* Lexington, MA: D. C. Heath & Company.

Sheller, M. (2014). 'The new mobilities paradigm for a live sociology'. *Current Sociology Review*, 62(6), 789–811.

Sheller, M. (2008). 'Mobility, freedom and public space'. In S. Bergmann & T. Sager (Eds.), *The ethics of mobilities: Rethinking place, exclusion, freedom and environment* (pp. 25–38). Aldershot, UK: Ashgate.

Sheller, M. & Urry, J. (2000). 'The city and the car'. *International Journal of Urban & Regional Research*, 24(4), 737–757.

Steinemann, A., Apgar, W. & Brown, H. (2005). *Microeconomics for public decisions.* Mason, OH: Thomson South-Western.

The Governor's Commission on the Los Angeles Riots. (1965). *Violence in the city – an end or a beginning?* Los Angeles: State of California.

Ureta, S. (2008). 'To move or not to move? Social exclusion, accessibility and daily mobility among the low-income population in Santiago, Chile'. *Mobilities*, 3(2), 269–289.

US Department of Transportation. (2004). 'National household travel survey, 2001: Summary of travel trends'. Retrieved from http://nhts.ornl.gov/2001/pub/STT.pdf.

Walzer, M. (1983). *Spheres of justice: A defense of pluralism and equality.* New York: Basic Books.

Wolff, J. (2008). 'Social justice and public policy: A view from political philosophy'. In G. Craig, T. Burchardt & D. Gordon (Eds.), *Social justice and public policy: Seeking fairness in diverse societies* (pp. 17–31). Bristol: The Policy Press.

Wright, E. O. (1997). *Class counts.* Cambridge: Cambridge University Press.

Zimmerman, M. (2012). 'Theorizing inequality: Comparative policy regimes, gender and everyday lives'. *The Sociological Quarterly*, 54(1), 66–80.

6 Mobile methods, epistemic justice and mobility justice

David Butz and Nancy Cook

Introduction

This chapter employs the concept of epistemic justice as a way to reflect on mobilities research in relation to mobility justice. We argue that epistemic justice is a significant aspect of mobility justice in which mobilities scholars are intimately involved through research practice. That is to say, researchers' methodological decisions have epistemic justice implications that matter for mobility justice. We develop this claim with reference to 'mobile methods', techniques that include embodied, kinetic and kinaesthetic researcher involvement in the mobile practices and contexts of the social groups or in the spaces under investigation. We focus on mobile methods because of their current prominence in mobilities scholarship and because the epistemological characteristics that are understood to recommend them for mobilities research present instructive challenges in terms of epistemic justice. Drawing from the socio-geographical context of our research in rural northern Pakistan, we outline four ways that mobile methods may contribute to epistemic injustice in certain research contexts: these relate to (a) the suggestion that mobile methods yield greater authenticity or accuracy of data, (b) the assumption that the participatory nature of mobile methods is inherently empowering to research subjects, (c) limitations mobile methods place on the categories of subject who can participate in mobilities research and the discretion they may exert while participating, and (d) the tendency of mobile methods to assume or construct an "overanimated mobile subject" (Bissell, 2010, p. 58).

Three caveats are required before we begin. First, these are propensities rather than inherent characteristics of mobile methods, as evidenced by many applications that avoid some of them. This chapter is not an indictment of mobile methods, but rather a discussion of susceptibilities. Second, they are propensities shared by numerous methods. Dubious claims to authenticity, accuracy and empowerment are common across qualitative research approaches, as are unjust sampling exclusions and skewed constructions of the research subject. Our intention is not to suggest that mobile methods are more likely to reproduce epistemic injustice than other techniques, but rather to focus on this currently popular approach to mobilities research as a way to demonstrate the connections among research practice, epistemic justice and mobility justice. Third, although we illustrate these propensities with reference to transcultural research in the rural Global South, we don't think they are distinctive to such contexts. They are neither inherent to mobile methods, nor merely the product of the contingencies of context.

Epistemic justice

Miranda Fricker (2007) identifies two forms of epistemic injustice. 'Testimonial injustice' occurs when "someone is wronged in their capacity as a giver of knowledge" (Fricker, 2007, p. 7); for example, when they are excluded from speaking or granted less credence as a knowledgeable speaker because of context-specific prejudice or structural arrangements that undermine their capacity to be received as credible. 'Hermeneutical injustice' transpires when "someone is wronged in their capacity as a subject of social understanding" (Fricker, 2007, p. 7) because "some significant area of one's social experience is obscured from collective understanding owing to hermeneutical marginalization" (Fricker, 2006, p. 102). It describes the failure of a discursive community to develop the vocabulary or concepts to render the experiences of some people intelligible.

If we accept Tim Cresswell's (2010) formulation that mobility is constituted though the entanglement of movement, representation and practice, then it follows that mobility justice as a value pertains not only to the characteristics of mobility regimes and the placement of social groups in mobility infrastructures and discourses, but also to individuals' capacities and opportunities to represent – to attach meaning to – the mobilities that shape their lifeworlds. Epistemic injustice impedes certain people's capacity to participate meaningfully in mobility's representational aspect. We think this is a mobility injustice in itself, as well as a contributor to other forms of mobility injustice. People whose mobility experiences and representations are prejudicially discredited, misunderstood or received as unintelligible are unlikely to be well-served by infrastructure planning, or by mobilities theorizing. One effect of linking epistemic justice with mobility justice in this way is to incorporate mobilities research into considerations of mobility justice; it raises the possibility that research practice may itself instantiate mobility injustice.

All social research risks testimonial injustice through processes that define study populations, select samples, choose research sites, etc. Researchers frequently make decisions that are sensible in the context of a specific project, but which exclude or devalue the testimony of some prospective subjects for reasons of convenience or prejudice. When whole bodies of knowledge are shaped by such credibility deficits and excesses, then the research process itself reproduces testimonial injustice, helping to create conditions for other axes of social injustice. Hermeneutical injustice is also an ever-present possibility because researchers' efforts to understand the articulated experience of research subjects are hampered by the mediating effects of our own preexisting vocabularies, conceptual frameworks and disciplinary conventions, which often impose meaning on them rather than accessing their self-intelligibility. Moreover, subjects' efforts to express themselves may be hampered by a lack of adequate vocabulary or conceptual resources to describe certain categories of experience, or by other power differentials inherent to specific research relationships that render them mute or unintelligible.

This is an epistemological and methodological issue as well as an ethical matter. We treat our research subjects unjustly – we wrong them "in their capacity as a subject of social understanding" (Fricker, 2007, p. 7) – if we fail adequately to recognize the epistemological characteristics of the knowledge performances they share with us; that is, if we are insensible to the power relations, interests and interactional contexts in relation to which subjects represent and perform mobilities to us or for us, and which shape their meaning. On the other hand, we nurture epistemic justice by developing methodologies that allow participants to represent and perform their mobilities in forms and contexts that enable intelligible self-expression and which – as part of that – foreground the interests, power relations and social contexts that shape the epistemological characteristics of those self-expressions and mobility performances.

Mobilities research in Shimshal, northern Pakistan

These matters preoccupy us in relation to our investigation of the mobility consequences of a newly constructed road for residents of Shimshal, an agricultural community of about 125 households located in the Karakoram Mountains of Pakistan's Gilgit-Baltistan administrative unit. When we first visited Shimshal in 1988, the settlement was three long days' walk from the Karakoram Highway, the region's only arterial road. The community had begun a challenging road-building project three years earlier, in 1985, relying mainly on local volunteer labor with intermittent support from government and international development agencies. It took another 15 years for the 60-kilometer jeep track linking Shimshal to the Karakoram Highway to be completed. When it finally opened to vehicular traffic in 2003, the three-day walk was reduced to a two-hour jeep ride, with dramatic and socially uneven implications for community life (Cook & Butz, 2011, 2018). Although the road's construction was only peripherally related to research we conducted during six field seasons in Shimshal between 1988 and 2003, it nevertheless became a *leitmotif* of our visits; we walked the footpath many times, drove completed sections of the road when we had the chance, listened to residents discuss the prospect of increased accessibility and observed in detail the construction process. In 2011, seven years after the first jeep arrived in Shimshal, we began a multi-year project to study Shimshalis' experiences of life after the road, turning to the mobility studies literature for guidance.

Mobile methods

Thus began our introduction to the 'mobilities turn' and its associated methodological literature. We learned that although various approaches have been employed and advocated to access the activities, events and meanings through which mobilities are socially instantiated, much emphasis has been placed on 'mobile methods', approaches used on the move to "capture, track, simulate, mimic, parallel and 'go along with'" subjects, objects, images and ideas in motion (Büscher, Urry & Witchger, 2010, p. 7; see also D'Andrea, Clolfi & Gray, 2011; Fincham, McGuiness & Murray, 2010; Hannam, Sheller & Urry, 2006; Larsen, Urry & Axhausen, 2006; Manderscheid, 2014; Spinney, 2015).

Explanations for this methodological focus stem from the claim that movement is a fleeting, materially embedded, embodied and experiential phenomenon, which cannot adequately be apprehended using cognition-oriented methods. Authors assert that accessing "the more intangible and ephemeral meanings of mobility" (Spinney, 2009, p. 826) requires moving with research subjects: participating in their movement experiences, eliciting their commentary in specific mobility contexts and observing practices and spaces of movement that would be inaccessible via verbal or textual representation alone. By enabling researchers to experience particular mobilities *in situ*, mobile methods are also understood to promote greater sensitivity to how the attributes of place shape and are constituted through movement (Hein, Evans & Jones, 2008). In addition, because mobile techniques commonly involve the embodied participation of research subjects, they are perceived to share epistemological and ethical advantages typically associated with participatory approaches.

Much of our previous research involved moving with Shimshalis in and through some of the spaces of their daily lives, sometimes as a deliberate accompaniment to standard ethnographic practices of interviewing and participant observation, and sometimes simply as a by-product of living in a pedestrian landscape with a local household for extended field seasons. We hadn't understood these strategies as mobile methods *per se*, but when we encountered the term we could appreciate the methodological benefits researchers attribute to techniques that involve "moving, being and seeing with" mobile subjects (Merriman, 2014, p. 174). At the same time, our previous field experiences suggested that such methods have epistemological limitations not widely acknowledged in the mobile methods scholarship. Accordingly, when designing the methodology for our new project, we opted to continue using the moving-with techniques we had previously employed, but combined them with other methods we thought would avoid some of these limitations. Among these were self-directed photography, mapping exercises with children and diagramming mobile family histories as well as documentary work, individual and group interviews and participant observation. Over the course of the project, as we reflected on the methodological preoccupations of the mobilities turn in relation to our research context, we developed a keener sense of the epistemological implications of a mobile methods approach and began to understand these in terms of epistemic justice and mobility justice. We turn now to four of these implications to exemplify the connections among research practice, epistemic justice and mobility justice.

Accuracy and authenticity

Advocates sometimes assert that mobile methods offer greater immediacy, closeness of engagement, authenticity of experience and, therefore, accuracy of data than desk-based and sedentary field-based techniques. Merriman (2014) identifies several problems with this epistemological progression – from researcher involvement and immersion, to enhanced phenomenological closeness to lived reality, to enriched authenticity of experience, to greater accuracy of results – including its potential to delegitimize conventional methods and stifle innovation, thereby undermining methodological pluralism and experimentation. One problem he doesn't address is that claims to authenticity and accuracy on the basis of experiential closeness and kinaesthetic involvement assume research subjects and their social environments are passively awaiting researchers' attention rather than actively and reflexively shaping the encounters researchers have with them. There is little recognition that mobile encounters and performances occur in a context of power relations, interests and interactional dynamics that undermine conventional authenticity, accuracy and closeness as epistemological attributes; that these are performative *effects* of kinaesthetic mobile experiences rather than characteristics inherent to them. Although mobile methods emphasize immersion, participation and responsiveness, they nevertheless risk treating research subjects as mere objects of knowledge, what Spivak (1999, p. 113) terms "Native informants", rather than acknowledging them as reflexive knowledge producers. The result is epistemic injustice: research subjects are misrecognized in their capacities both as givers of knowledge and as subjects of social understanding.

As ethnographers working in the formerly colonized and currently marginal and peripheral borderlands between Central and South Asia, we are keenly aware that we interact with locals in an asymmetrically constituted transcultural field of power/knowledge and that the performances, representations and interactions that comprise our 'data' reflect this discursive field in often indeterminate ways that preclude straightforward notions of accuracy or authenticity. As an alternative to such epistemological attributions, we draw from Mary Louise Pratt's (1992, 1994) analysis of transcultural interactions in colonial contexts to conceive our research field as 'authoethnographic'; that is, comprised of reflexive – even self-interested and interventionary – self-performances, self-representations and self-actualizations that produce rather than reflect a version of truth (Butz, 2010; Butz & Besio, 2004). It is not epistemologically sound to treat the instantiations of mobility we encounter or experience in Shimshal as capturing or accessing realities about the road's significance that preexist their visual/narrative/performative expression to us. Instead, we must read them as transcultural performances of identity in relation to the road, which have been shaped by the affordances and constraints of our methodology and by the field of power in which it was employed.

From this perspective, our obligation is not to chase the *chimera* of accuracy or authenticity, but rather to develop approaches that enable relatively unrestricted opportunities for autoethnographic expression by our research subjects and constitute data in a form that is amenable to interpretation and analysis as autoethnographic. This is an important step toward epistemic justice: understanding the social worlds we encounter as populated by knowledgeable agents who have the capacity and right to be understood in relation to their own experiences and processes of self-constitution. There is no reason why mobile methods cannot be coupled with an epistemology of authoethnography, rather than an epistemology of accuracy, to achieve this ethical benefit; indeed, they may be well-suited to such an epistemological sensibility. But that requires abandoning the notion that kinaesthetic immersion in a particular context of mobility offers special access to authenticity of experience or accuracy of apprehension.

To give a sense of how constraining an attachment to accuracy and authenticity is to understanding how the road is (re)shaping Shimshalis' lifeworlds, we turn to a photo/narrative combination produced by elderly farmer Barkat Ali Shah as part of the self-directed photography component of our research. In response to the generic question "what is the significance of the Shimshal Road for your everyday life?", Mr. Shah produced a carefully arranged photo of several "generations of lighting" (Figure 6.1). The photo is static – almost a still life – but it became animated through Mr. Shah's explanation, which associated each type of lighting (grease lamp, coal oil lantern, hurricane lamp, high-pressure propane lamp, lightbulb) with a different period in Shimshal's history of mobility. He included detailed descriptions of himself and his father carrying lamps and other unwieldy items to Shimshal by foot and emphasized the transportation possibilities afforded by successively improved footpaths culminating with the jeep road. He described his family as lighting innovators in Shimshal, while also reminding research assistants of the mobility difficulties their ancestors faced. He enrolled us as allies in this didactic and self-representational effort because, unlike our youthful assistants, we had walked the route to Shimshal many times before the road's completion and remember when currently obsolete forms of lighting were common.

The association Mr. Shah constructed between lighting and mobility infrastructure, and the way he wove his family's achievements into that association, alerted us to connections Shimshalis draw between mobility and identity, and to a corner of his life in which he discerns the road's influence. To search in this deliberately composed photo or its accompanying narrative for an accurate or authentic glimpse of a preexisting reality would be to miss their epistemological potential for understanding the road (and method) as a resource for Mr. Shah to formulate a sense of himself in relation to his

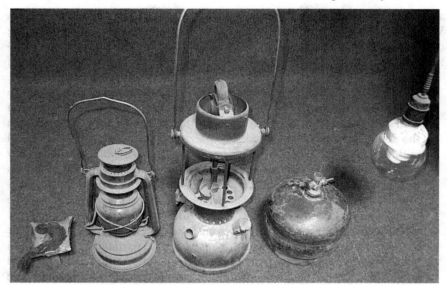

Figure 6.1 Generations of lighting
Source: Photograph by Barkat Ali Shah, 59, male, farmer/animal health specialist.

community and its material environment. It would also instantiate an epistemic injustice by negating the credibility of his self-performance and refusing to recognize the terrain of his own self-intelligibility.

Participation and empowerment

A second potential weakness of mobile methods in terms of epistemic justice is a tendency to assume too direct a connection between participation and empowerment. Mobile methods are participatory in two senses: they allow researchers to participate kinaesthetically in research subjects' movement experiences; and they often involve subjects actively in the research process through various types of participant-led go-alongs. Participation in the second sense is often understood as ceding representational and decision-making control to participants in ways that empower them (Anderson, 2004; Evans & Jones, 2011). According to that logic, empowerment becomes an ethical basis for participatory mobile methods.

Although techniques that solicit subjects' commentary and observe their actions on-the-move have epistemological benefits, especially in relation to understanding the spatial constitution of knowledge, meaning and behavior (Evans & Jones, 2011), to frame those benefits in terms of empowerment misrecognizes the epistemological basis of subjects' self-performances in those contexts and risks constituting an epistemic injustice. Our doubts about the participation-as-empowerment narrative have two aspects.

First, in many circumstances research subjects may feel more beleaguered than empowered by having decision-making responsibility thrust upon them as they move through a landscape with an unfamiliar researcher. Saskia Warren (2017, p. 796) articulates the crux of this reservation while reflecting on her use of walking interviews to consider Muslim women migrants' practices and experiences of public space in Birmingham in the United Kingdom:

> In fact this methodological causality – self-determination leading to greater empowerment – should be reconsidered when it is held in tension with cultural and social difference, and the hierarchies of power between the researcher and researched. A deferential approach to the cultural authority of "the teacher" (in this case a university researcher) was a key contributing factor that frustrated attempts to realign power relations through research design in the walking interview. Further inter-generational asymmetry in power was evident between the participants: the older female led; the younger female followed. Participants regularly asked for reassurance that their selected route was suitable, and whether the researcher would like to visit anywhere in particular. Rather than "empowering" the participants, the go-along scenario in this case proved challenging. It pushed beyond comfort zones where leading the route meant unwanted authority was assumed, exacerbating any unease in what is already an unusual socio-spatial situation.

Warren's assessment resonates with our research experience and illustrates how mobile methods may inhibit research subjects' capacity to represent and perform mobilities to their satisfaction. Moreover, if the empirical yield of walking interviews conducted in circumstances like those described by Warren is analyzed as deriving from a context in which interviewees experienced empowerment, then the conclusions drawn would likely be incommensurate with interviewees' own understanding of the mobilities in question.

Second, many development studies scholars reject the notion of empowerment through participatory methods on theoretical and political grounds, associating it with neoliberal ideologies of responsibilization and crude theories of power and questioning assumptions that participatory methods enable more truthful or agential expression. Uma Kothari (2001, p. 140), for example, argues that to imagine empowerment is achieved by ceding representational control supposes

> that people who wield power are located at institutional centers, while those who are subjected to power are to be found at the local or regional level – hence the valorization of "local knowledge" and the continued belief in the empowerment of "local" people through participation.

A discourse of empowerment through participation overlooks how power relations organize the local and how participatory techniques can articulate local knowledge in ways that reproduce local power structures. Members of subordinated groups may feel compelled to articulate norms that reassert established hierarchies, thereby legitimizing local knowledge in ways that perpetuate their subjugation.

Moreover, the power researchers exercise in research encounters and on local self-representations is neither discrete enough nor sufficiently located in our own agency to be within our power to relinquish or redistribute. We cannot expunge our influence, or that of powerful local actors and discourses, from subjects' self-representations or offer them speaking or performative positions outside the field of power through which they are constituted. To understand participatory methods as empowering is a misreading of the epistemological characteristics of the self-representations and self-performances we receive from participants.

Empowerment is a less plausible ethical criterion for assessing mobile and other participatory methods than is epistemic justice, which can be nurtured by employing methods that individuals from a variety of subject positionings can engage with willingly, comfortably and competently, and by adopting an epistemological stance that strives to account for the influence of internal and transcultural power dynamics on research participants' self-representations and performances. Another photo/narrative combination produced in our self-directed photography project illustrates this distinction between empowerment and epistemic justice.

Community activist Zul Haja used the photo "Nazia's umbrella" to argue that Shimshalis' identities have recently become shaped by integration with wider currents of thought and knowledge (Figure 6.2). She related that more people are using umbrellas, sunscreen and wide-brimmed hats; villagers have become more aware of the damaging health effects of solar radiation and are especially concerned to protect their children from the sun (note the small child on Nazia's back). She speculated that as Shimshali women become more exposed to South Asian and global ideologies of youthful beauty, they are motivated to retain a fair, unwrinkled complexion. In Ms. Haja's telling, Nazia's umbrella symbolizes a new type of Shimshali femininity: educated and preoccupied with downcountry standards of beauty and motherly responsibility. As one of Shimshal's few female community leaders, and a self-styled role model for young women, she used her photo to promote a version of femininity that she aspires to exemplify. Although the road figures implicitly in her reference to imported cultural influences, it serves mainly as a device for furthering her own identity project.

It might be tempting to understand Ms. Haja's self-confident, self-interested and premeditated account of changing female gender identities in terms of empowerment, and she was no doubt able to exercise greater control over her self-performance through self-directed photography than a go-along mobile method would have allowed. However, it is also clear that her narrative is shaped by a variety of power relations that were not erased or submerged by our participatory method. These include the influence of metropolitan discourses on her framing of feminine identity, the imposition of her own modern and educated subject positioning on Shimshali women in general and

Figure 6.2 Nazia's umbrella
Source: Photograph by Zul Haja, 35, female, homemaker/local council member.

her efforts to impress us with her knowledge of atmospheric science and downcountry beauty standards. It is plausible to say that Ms. Haja used self-directed photography to express herself credibly, intelligibly and subtly in an autoethnographic register without claiming that the power relations evident in her self-performance amount to empowerment.

Sampling and subject self-expression

Mobile methods involve two types of sampling decisions: who to accompany, observe or converse with on the move; and in what social, spatial or temporal contexts. In many situations both sets of sampling options are constrained by the go-along nature of the method in relation to the social context under investigation, in ways that exclude the mobility-inflected spaces and experiences of some prospective subjects. To the extent that they (or their spaces of mobility) are excluded from a sample on the basis of methodologically determined possibility or convenience, rather than the relevance of their experience or credibility of their testimony, they are unjustly denied the opportunity to have their perspectives and experiences considered.

Our research context offers a good example of the systematically biased sampling constraints a go-along method may impose. Shimshal has many important spaces of mobility that researchers like us simply cannot have embodied access to, with or without local accompaniment, for reasons of privacy and propriety. These include certain domestic and religious spaces, and spaces that are gendered in particular ways. A participatory mobile method would not give us embodied access to these spaces or the practices that occur there, even though locals are free to tell us about these spaces and things that happen there or to show us photographs. Certain other village spaces are accessible to us, but not at certain times or with certain categories of local subject, whose reputation would be jeopardized by being co-present there with a foreigner. Young and middle-aged women especially have to be careful about what spaces they occupy, with whom and on what business, but other groups in Shimshal face similar constraints. Not surprisingly, moving-with methods would mostly limit the participation of the groups – women, children, poor people – whose perspectives and experiences are already underrepresented in research on the area, thereby reinforcing existing testimonial injustice.

An additional concern is subjects' limited capacity to shape researchers' moving-with experience during a go-along and, therefore, to be understood on their own terms and at their own discretion. Subjects may have liberty to select a route, determine the pace of movement and comment on selected features or activities, but otherwise they have little control over researchers' encounters with the mobility context or practices in question. This may seem like an epistemological advantage to researchers, whose apprehension of events, practices or spaces is less intensely mediated by the interests and pre-occupations of research subjects than would be the case with interviewing or photo-elicitation, for example. But that comes at a cost to participants, who may wish to present or represent themselves and their environments in more selective and bounded ways. In other words, kinaesthetic immersiveness may actually *reduce* participants' capacity to express themselves on their own terms, thereby undermining a key value of epistemic justice. Ironically, it may also reduce interpretive richness. An example from our self-directed photography project helps to illustrate these points.

Teenaged student Naveed Ahmed addressed the road's significance by photographing an assortment of new electric kitchen appliances located in a relative's traditional kitchen (Figure 6.3). The picture portrays the material makeup of a household space and hints at aspects of women's reproductive labor. His accompanying commentary reflected on household economies and changes in how marital partners relate. Mr. Ahmed stated that the man in this household works long hours for low wages in a distant city to purchase and transport consumer items that save his wife a small amount of labor. In his telling, these appliances have little use in Shimshal, except to signify the man's appreciation for his wife's efforts to maintain a home in his absence or as a symbol of purchasing power. It is hard to imagine this intense focus on a micro-environment or Mr. Ahmed's sophisticated commentary arising from a go-along method in which selectivity and premeditation are more difficult. Eager to perform studiousness, he carefully selected his scene, composed his picture and spoke with the homemaker before offering commentary. He could

Figure 6.3 Old kitchen with modern equipment
Source: Photograph by Naveed Ahmed, 18, male, student.

have taken us to this house, showed us the appliances and talked about them *in situ*, but not without exceeding the bounds of propriety or exposing us to other parts of the house, other artifacts or household members, any of which may have interfered with his argument and prevented him from shaping his intervention as he wished. A go-along visit to the kitchen may have given us a feeling of kinaesthetic closeness to the space in question, with commensurate illusions of accuracy and authenticity, but with some cost to Mr. Ahmed's representational discretion and at the loss of a thought-provoking 'insider' analysis borne of self-conscious premeditation.

An "overanimated mobile subject"

It is said that from the perspective of a hammer, everything looks like a nail. Similarly, from the vantage of mobile methods, the complex socio-spatial relations that constitute mobilities may seem to be encapsulated most significantly by those phenomena mobile methods are designed to access – experiences, practices and performances of subjects *in motion* – resulting in the assumption of "an *overanimated* mobile subject" (Bissell, 2010, p. 58, quoted in Merriman, 2014, p. 177; emphasis in original). Despite the fact that "stillness, waiting, slowness and boredom may be just as important to many situations, practices and movements as sensations and experiences of speed, movement, excitement and exhilaration", mobile methods "do not seem to be very good at registering the more passive practices, engagements and affective relations which gather around movements and mobilities" (Merriman, 2014, p. 177). Neither do they attend effectively to comparatively immobile material artifacts and structures, the characteristics of which may be important manifestations of mobile practices and infrastructures.

Such systematic partiality of focus, which has ontological dimensions, creates conditions for testimonial and hermeneutical injustice: it excludes the experiences of subjects whose relations to mobility are less animated than others and reduces the possibility for certain types of mobility-related experience or practice to be articulated or apprehended intelligibly. In Shimshal, for example, the embodied mobilities of much of the population have not been noticeably affected by the road's construction, but all villagers' lives have been reshaped by a new landscape of artifacts transported along the road. The road's 'meaning' to many Shimshalis relates less to greater ease of personal travel than to changes in their houses, large-scale re-landscaping of the village and increasing access to information *in situ*. Mobile methods would provide little insight into these implications of mobility regime change.

Moreover, many adult women and elderly villagers have never traveled the Shimshal Road or even ridden a wheeled vehicle, and for many others jeep travel is still a rare experience. Such people practice pedestrian mobilities, which interface to some extent with road-based movements; but focusing on the embodied and experienced performance of these mobilities would provide an impoverished picture of the road's significance for Shimshalis' everyday lives, which, for many, relates more to the cement construction of a new dwelling, improved educational facilities or increased availability of consumer goods in village shops. And focusing on the road-based mobilities of those people who do use the road in an active embodied way would reproduce the disproportionate influence of men, and women from prosperous families, on the production of mobilities knowledge.

To convey the importance of relatively fixed objects to Shimshalis' experience of road-oriented mobility, we turn to a photo/narrative pairing produced by Zaib Aman, a woman from a comparatively prosperous family (Figure 6.4). In her telling, this "modern washroom" would have been impossible to construct in Shimshal prior to the road's completion; it is therefore a space where her family experiences the benefits of vehicular mobility. She emphasized that just a few wealthy households can afford modern bathrooms, but everyone in Shimshal has become more concerned about cleanliness and covets these facilities. Ms. Aman used the picture to construct herself as modern, wealthy and well educated – a trendsetter – in keeping with her family's social position. The photo's contents and accompanying narrative perform her identity as distinct from other Shimshalis, while also framing the community as aspiring to the hygienic modernity she depicts. Although Ms. Aman is one of the few Shimshali women who travels the road frequently, she did not highlight her personal mobility to signify her family's prosperity and modernity, but rather focused on

Figure 6.4 Modern washroom
Source: Photograph by Zaib Aman, 48, female, homemaker.

household fixtures whose existence in Shimshal depend on vehicular mobility. Ms. Aman's photo/narrative cannot be imagined to provide richer or more authentic insight than mobile methods would achieve, but it does illustrate the sorts of mobility experiences and discourses that may be overlooked by such approaches, even though they are significant to the people whose mobilities we wish to understand.

Conclusion

Our objective has been to use the example of mobile methods to demonstrate that methodological decisions mobilities researchers make have epistemic justice implications that matter for mobility justice. Using our research context in northern Pakistan as an illustrative case, we have argued that mobile methods risk testimonial and hermeneutical injustice by: (a) adopting an epistemology of accuracy and authenticity that misrecognizes the autoethnographic character of human-to-human research encounters; (b) conceiving the participatory aspects of mobile techniques as empowering to research subjects, thereby underestimating the extent to which the specificities of method, the relative positioning of researchers and subjects in power/knowledge circuits and local power structures may constrain subjects' self-expression and mobile self-performances; (c) placing methodologically determined limitations on whose mobilities can be examined in what spatio-social context, and on research subjects' discretion to shape researchers' apprehension of their mobilities; and (d) focusing narrowly on experiences, practices and performances of subjects in motion to the exclusion of individuals and social groups whose relation to mobility may be shaped more by artifacts, spaces and other people's movements than by their own. The foregoing discussion establishes a link between epistemic injustice and some characteristics commonly associated with mobile methods. It remains in our final paragraphs to trace how these methodological instantiations of epistemic injustice constitute mobility injustice.

First, starting from the notion that mobilities are constituted through the entanglement of physical movements, experienced practices and representations of movement (Cresswell, 2010), and borrowing from Iris Marion Young's (2011) conception of social justice, it follows that individuals or social groups are victims of mobility injustice if they are deprived of the means to develop and exercise their capacities in any of these mobility dimensions at the same time as the social processes that are responsible for their deprivation enable others to develop or exercise their capacities in relation to the same dimensions. Testimonial and hermeneutical injustice describe two forms of mobility deprivation in the realm of representation. By excluding or limiting some subjects in their capacity as givers of mobilities knowledge and subjects of social understanding regarding their own mobility practices and experiences, and by focusing disproportionately on certain highly animated features of mobility constellations, practitioners of mobile methods risk producing theorizations and accounts of mobility that compound representational deprivation and, therefore, mobility injustice for some social groups.

Second, as we have shown, go-along methods may have the effect of regulating subjects' mobility performances, narrations and experiences in real time. This constitutes governance of subjects' mobility through researchers' mobility, whereby researchers' discretion to move with research subjects limits the latters' capacity to perform or represent their mobilities to researchers to their satisfaction. In at least some circumstances this constitutes an unjust constraint on mobility self-determination, the significance of which is compounded when researchers attribute the epistemological and ethical values of authenticity, accuracy and empowerment to the disciplined mobility performances that result. That is, a relatively insignificant momentary performative and representational deprivation becomes a more significant denial of a research participant's capacity as a giver of knowledge and subject of social understanding when it occurs in the context of a research process where the effects of that deprivation are unrecognized or misconstrued analytically in the production of academic knowledge.

Acknowledgments

The research that inspired this chapter was supported by Canada's Social Sciences and Humanities Research Council [grant number 410 2009 0579] and the Brock University Council for Research in the Social Sciences. We extend sincere thanks to the villagers of Shimshal, whose desire for road connectivity motivated our interest in mobility and whose friendship and cooperation have made our research in the community a privilege and a pleasure. Parts of this chapter are reproduced in altered form, with permission, from Butz & Cook (2017).

References

Anderson, J. (2004). 'Talking whilst walking: A geographical archaeology of knowledge'. *Area*, 36(3), 254–261.

Bissell, D. (2010). 'Narrating mobile methodologies: Active and passive empiricisms'. In B. Fincham, M. McGuinness & L. Murray (Eds.), *Mobile methodologies* (pp. 53–68). Basingstoke, UK: Palgrave Macmillan.

Büscher, M., Urry, J. & Witchger, K. (2010). *Mobile methods*. London: Routledge.

Butz, D. (2010). 'Autoethnography as sensibility'. In D. DeLyser, S. Herbert, S. Aiken, M. Crang & L. McDowell (Eds.), *The Sage handbook of qualitative geography* (pp. 138–155). London: Sage.

Butz, D. & Besio, K. (2004). 'The value of autoethnography for field research in transcultural settings'. *The Professional Geographer*, 56(3), 350–360.

Butz, D. & Cook, N. (2017). 'The epistemological and ethical value of autophotography for mobilities research in transcultural contexts'. *Studies in Social Justice*, 11 (2), 238–274.

Cook, N. & Butz, D. (2018). 'Gendered mobilities in the making: Moving from a pedestrian to vehicular mobility landscape in Shimshal, Pakistan'. *Social & Cultural Geography*, 19(5), 606–625.

Cook, N. & Butz, D. (2011). 'Narratives of accessibility and social change in Shimshal, northern Pakistan'. *Mountain Research and Development*, 31(1), 27–34.

Cresswell, T. (2010). 'Towards a politics of mobility'. *Environment and Planning D: Society and Space*, 28(1), 17–31.

D'Andrea, A., Ciolfi, L. & Gray, B. (2011). 'Methodological challenges and innovations in mobilities research'. *Mobilities*, 6(2), 149–160.

Evans, J. & Jones, P. (2011). 'The walking interview: Methodology, mobility and place'. *Applied Geography*, 31(2), 849–858.

Fincham, B., McGuiness, M. & Murray, L. (2010). *Mobile methodologies*. New York: Palgrave Macmillan.

Fricker, M. (2007). *Epistemic injustice: Power and the ethics of knowing*. New York: Oxford University Press.

Fricker, M. (2006). 'Powerlessness and social interpretation'. *Episteme: A Journal of Social Interpretation*, 3(1–2), 96–108.

Hannam, K., Sheller, M. & Urry, J. (2006). 'Editorial: Mobilities, immobilities and moorings'. *Mobilities*, 1(1), 1–22.

Hein, J., Evans, J. & Jones, P. (2008). 'Mobile methodologies: Theory, technology and practice'. *Geography Compass*, 2(5), 1266–1285.

Kothari, U. (2001). 'Power, knowledge and social control in participatory development'. In B. Cooke & U. Kothari (Eds.), *Participation: The new tyranny* (pp. 139–152). London: Zed Books.

Larsen, J., Urry, J. & Axhausen, K. (2006). *Mobilities, networks, geographies*. Aldershot, UK: Ashgate.

Manderscheid, K. (2014). 'Criticizing the solitary mobile subject: Researching relational mobilities and reflecting on mobile methods'. *Mobilities*, 9(2), 188–219.

Merriman, P. (2014). 'Rethinking mobile methods'. *Mobilities*, 9(2), 167–187.

Pratt, M. L. (1994). 'Transculturation and autoethnography: Peru, 1615/1980'. In F. Barker, P. Hulme & M. Iversen (Eds.), *Colonial discourse/postcolonial theory* (pp. 24–46). Manchester, UK: Manchester University Press.

Pratt, M. L. (1992). *Imperial eyes: Travel writing and transculturation*. London: Routledge.

Spinney, J. (2015). 'Close encounters? Mobile methods, (post)phenomenology and affect'. *Cultural Geographies*, 22(2), 231–246.

Spinney, J. (2009). 'Cycling the city: Movement, meaning and method'. *Geography Compass*, 3(2), 817–835.

Spivak, G. (1999). *A critique of postcolonial reason*. Cambridge, MA: Harvard University Press.

Warren, S. (2017). 'Pluralizing the walking interview: Researching (im)mobilities with Muslim women'. *Social & Cultural Geography*, 18(6), 786–807.

Young, I. M. (2011). *Responsibility for justice*. Oxford: Oxford University Press.

Justice and mobility infrastructures

7 The autonomobility system
Mobility justice and freedom under sustainability

Noel Cass and Katharina Manderscheid

Introduction

In this chapter we critique automobility, the hegemonic mobility system of the present, and suggest an alternative imaginary we call 'autonomobility'. This utopian future mobility system aims at reconciling mobility justice requirements, environmental restrictions on mobility and, distinctively, autonomy as the supervening value: a freedom of self-determination. While previous distributive justice models address equity in spatial access to life opportunities, autonomobility more clearly reconfigures mobility justice to acknowledge that freedom from compulsion (to stay or to move) is the value required to equitably ensure mobility justice under scarcity. We offer histor- ical, scenario-based, fictional and existing glimpses of autonomobility to suggest it is a viable model of pegging the pursuit of individual flourishing within environmental limits and distributional equity, using socialized and collectivized forms of co-movement as a vision to animate an egalitarian society.

Automobility and its environmental and justice problems

The "'system' of automobility" (Urry, 2004) defines personal transport and the ordering of mobility, space and society in the modern age. It comprises the steel-and-petroleum car itself, its production, fuel and infrastructure industries, the policies that create automobile landscapes that separate work, residence and other activities in space, and the discursive association of cars with freedom and autonomy (Paterson, 2007). Modern lifestyles centered on suburban one-family houses and shopping and leisure facilities on the edge of cities represent the 'good life' under automobility.[1] Environmental crisis, 'peak oil' (Aftabuzzaman & Mazloumi, 2011) and potentially 'peak car' (Goodwin & Van Dender, 2013) constitute sound reasons to think beyond automobility (Urry, 2013), but critiques of it extend beyond envir- onmentalism. Cars and their infrastructures demand huge amounts of space and cause numerous fatalities through accidents and air pollution (Zhang & Batterman, 2013). Moreover, this mobility regime has severe social justice implications. As environmental justice (Havard, Reich, Bean & Chaix, 2011; Mitchell & Dorling, 2003) and transport inequality studies (Lucas, 2004) show, the social groups most affected by traffic emissions and noise are the socially deprived: automobility compounds other injustices. Housing markets also force poorer households to live on urban peripheries poorly connected to public transport, necessitating cars to access work, shops and other social infrastructures and to maintain social relations (Mullen & Marsden, 2018). Automobility unjustly imposes environmental and exclusionary 'bads' on nondrivers and compulsions to be mobile on drivers (Ureta, 2008).

To tackle these problems, sustainable transport debates suggest environmental impacts have primarily technological solutions (e.g., electrification), while others propose that accidents will be tackled by a shift from human to technological agency in autonomous vehicles (Manderscheid, 2018). Such 'technical' solutions ignore systemic issues, the increasing compulsions to travel, social injustices and freedom constraints in the automobile-centered mobility system (Better Transport, 2017). Some distributional justice issues are addressed through discourses of car dependence (Mattioli, Anable & Vrotsou, 2016), transport poverty (Simcock & Mullen, 2016) and mobility-based social exclusion (Delbosc & Currie, 2011). However, these discourses still often unproblematically equate (auto)mobility with social inclusion or opportunity, replicating associations between moving, freedom and justice in the sense of access to social 'goods'.

This equation makes sense in the classic model of transport justice as a matter of distributive principles (Golub & Martens, 2014; Martens, 2012, 2016). In this Walzerian model, the state, primarily, equitably distributes a transport 'good', which appears strangely abstract and non-relational, despite its deployment by sustainable mobility authors with a fully relational outlook, and seems to flip between being "accessibility as a human capability [and ...] accessibility to key destinations" (Pereira, Schwanen & Banister, 2017, p. 170). The human aspect of 'accessibility' is a potential to be mobile in order to access goods, services and opportunities, and site accessibility is their capability to be accessed. With origins linked to (transport) planning, this somewhat top-down, mechanistic model assumes that mobility and the potential to be mobile emerge from transport access and distribution, which will occur from above through investment in fixed systems based on areas and responding to assumed long-term 'needs' linked to social groups and geography rather than individuals. Daniel Newman (2017) frames the same issue in the terminology of rights, seeing the mobility required to access goods and services as conferring a right to affordable, environmental public transport, comparable to the right to adequate housing. Again, this takes the *need* for mobility as a fixed requirement of the good life without questioning the underlying compulsions to be (auto)mobile (Urry, 2002).

Transport as a 'good' is similar to 'travel demand' in being primarily derived, dependent on the meaning and importance of the other goods and services to which it provides 'access'. Socio-spatial inclusion or exclusion is not predictable *a priori*, being "an emergent property of the interaction between social practice and obligation, individual resources and physical infrastructure" (Cass, Shove & Urry, 2003, p. 28). Our point of departure is thus problematizing compulsions to be mobile in order to reconcile the connected environmental and justice dimensions of 'sustainable' mobility. We establish that the current (auto)mobility system is unsustainable and unjust, and curtails freedoms, and then construct an anchoring vision of a freedom-focused, environmentally sustainable *and* socially just mobility regime – 'autonomobility' – building on circulating scenarios of future mobility systems. Rather than taking access to mobility or certain sites as an end (as proposed in a distributive model of mobility justice), autonomobility entails both a right to move and a right not to. In a final step we identify (pre)existing seeds of the autonomobility system in history, fiction and present mobility practices that prefigure possible ways ahead (Gordon, 2008), whilst acknowledging obstacles and counter-forces to this vision. This approach moves beyond neutral academic analysis as 'transformative research' aiming to tackle the grand challenges of society (Schneidewind, 2015).

Automobility and its present economic and cultural foundations

Understanding automobility as a *system* helps explain its stability and persistence. The car industry plays a central role in capitalist accumulation even after Fordism (Aglietta, 2000), and there is a strong correlation between economic growth and the growth of motorized transport and traffic (OECD, 2000). Such capitalist accumulation regimes need to be complemented by modes of regulation; a sphere beyond the economy involving social values, understandings of the good life and constructions of 'normality' (Aglietta, 2000; Manderscheid, 2014). In the case of automobility, its discursive and cultural foundations draw on the promise of mobility as freedom, individuality and independence (Rajan, 2007). While historically mobility was "associated with insecurity and danger ..., in modernity, ... [it] gradually turned into a *common right* claimed among equals" (Rammler, 2008, p. 61). Continuous and linked increases in progress, freedom, welfare and also mobility remain the central promise of modernity. Nicolas Rose (1996, 1999) claims this modern conception of freedom and its strong link with mobility is of a very specific, individualized form, which he understands as a Foucauldian disciplinary or governmental technology, placing responsibility for social and economic wellbeing increasingly onto the neoliberal individual.

In labor markets, opportunities for wellbeing (distributed through employment) are not evenly distributed geographically. The modernist project seeks to smooth the disparities between employment opportunities and skills shortages with large-scale transport infrastructures and also with the increase of labor mobility (Bonin et al., 2008; Harvey, 2001). Thus, the modern promise of mobility and freedom, on which distributive models of mobility justice also rest, primarily means the mobility of economically productive labor forces. This becomes visible when looking at moral panics, which have often centered on "bad movers" (Urry, 2007, p. 205): the 'masterless men' of 17th-century England, gypsies across

Europe for 500 years and, more recently, 'New Age travellers', migrants and refugees. This capitalist labor-mobility regime currently shows itself in the strengthening of border controls across and around Europe to manage and filter labor mobility, variably requiring cheap labor or skills. Phrasing it drastically: only the economically useful, in set amounts, residing in predictable areas and at the command of capital are supposed to move (Huxley, 2006; Peters, 2006).

Also relevant to freedoms and compulsions, mobilities imply, rest on and enforce immobilities, such as those of the housing estate bisected by a motorway, the service station worker or the house-spouse doing housework and care work. But compulsions apply to the mobile as well as the immobilized. Employers expect employees to move, backed up by the threat of replacement (Schor, 2014), and welfare policies require unemployed people to accept jobs involving significant commuting distances or relocation for work, backed up by threats of sanctions.[2] The mobility of some – as well as the immobility of others – is a collective strategy or compulsion embedded in economic and material structures (Castells, 2002; Graham & Marvin, 2001), shaped by political decisions and path dependencies, constituted and stabilized by the production of corresponding discourses and subjectivities (Manderscheid, 2014; Paterson, 2007; Peters, 2006).

Against these 'dark sides' of automobility – its negative ecological impacts, social selectivity, injustices and compulsions – we now sketch a positive imaginary of a future mobility system. This system must address operating within environmental limits, the inequitable distribution of the life chances to which mobility provides 'access' and unjust compulsions either to move or stay immobile. Such an imaginary of socially and ecologically sustainable transport could help negotiate future ways of life and mobility.

Autonomobility as part of an imaginary of future sustainability

Others have already imagined alternative mobility futures 'beyond the car'. Dennis and Urry (2009, pp. 131–164), for example, outline three potential scenarios for 2050. The first two raise serious problems for justice and freedom. Their "regional warlordism" scenario (Dennis & Urry, 2009, pp. 151–155) is a dystopia in which mobility, energy and communication infrastructures collapse and long-distance mobility is only available for the superrich, who otherwise shelter in gated communities. Aspects of the decline or 'splintering' (Graham & Marvin, 2001) of infrastructures in socio-spatial peripheries and the disconnection of unproductive communities from social resources, alongside films such as *Children of Men* or *Mad Max*, hint at futures based on such extreme inequalities and injustices. With "digital networks of control" (Dennis & Urry, 2009, pp. 155–160), they imagine a high-tech, centrally coordinated surveillance-traffic landscape of what would now be called 'autonomous vehicles': a 'smart' but cocooned system of individual mobility. The scenario limits the freedom to move 'uncontrolled' by smart technologies as well as retaining the environmental implications of embedded carbon in single-occupant vehicles and smart technology (Wadud, MacKenzie & Leiby, 2016).[3]

Both scenarios threaten and reduce mobility, liberties and justice, involve inequitable distributions of scarce resources and include violence or a panoptic mobility regime. They also assume the continuation in some form of existing socioeconomic, political and mobility systems. Only Dennis and Urry's third scenario of "local sustainability" (2009, pp. 149–151) assumes a challenge to a capitalist foundation, involving economic downscaling, downsizing and degrowth (D'Alisa, Demaria & Kallis, 2014). Harvey (2000, p. 270 ff.) and North (2010) similarly imagine transportation becoming free, slow and primarily local alongside decentralized production and local social networks. These scenarios are built on the premises of restructured and de-globalized economic activities, whether this is produced exogenously through scarcity and the breakdown of unsustainable global chains or chosen adaptively to build social resilience to such threats. Yet these radical models of sustainability are seen either as wishful dreams (Harvey, 2000) or as unrealistic environmentalist discourse in which a 'contraction' of society would require "huge reversals of almost all the systems of the twentieth century, as well as … massive restructuring of economic activities and the displacement of the global organization of economy, finance and social life" (Dennis & Urry, 2009, p. 151).

Elements of autonomobility

Nevertheless, we wish to extend this future imaginary (ultimately only as unrealistic as the continuation of the status quo) by embedding further ethical values into the mobility systems of radical environmental sustainability. This requires prioritizing autonomy, sustainability and justice, localization and its implications, and collective mobility.

Autonomy, sustainability and justice

Until now we have refrained from a detailed discussion of justice and sustainability. Both have general and transport/mobility-related meanings. Sustainability consists of an orientation towards the future and the imperative that practices, structures and modes of living must not reduce the opportunities and chances of future generations. Whereas the environmental 'leg' of sustainability (Newport, Chesnes & Lindner, 2003) is typically defined through climate change research, *social* sustainability cannot be defined 'from outside'. Rather, the acceptability of socially unequal chances and opportunities is a matter of intra-social debate (Mouffe, 2011). Both sustainability and social justice, however, can be said to rest on inter- and intra-generational justice issues: a relationality in time and space. Thus, when we speak of just or sustainable (mobility) systems, we refer to this *relationality* underlying both concepts. With this working definition, we want to highlight problems stemming from an understanding of transport justice in terms of accessibility and distribution, which rests on an individualistic distributive approach that obscures these present and future relationalities.

We assert *freedom* as a central value of an autonomobility system, understood not in 'positive' terms of mobility being individually unlimited (or, rather, only limited by economic power), but 'negatively' as *freedom from compulsion* (Berlin, 1969). Translated into common-sense terminologies of *rights*, there should be, as far as possible (with the limits defined by present and future impacts), a right to move and a correlative right to remain immobile if necessary/desired. Operating under conditions of scarcity, such rights would not be absolute, but relative, and judged in the contexts of social necessity and value, subject to social negotiation. However, they must also operate within the importance given to maximizing individual flourishing (Nussbaum, 2001). In other words, a corresponding understanding of 'autonomy' (rather than freedom) rests heavily on a philosophical switch from ideas of distributional justice based on utility, subjective welfare or access to certain resources towards Sen's priorities of pursuing the accomplishment of functionings for all people in society equitably by maximizing their capabilities. The 'capability approach' has been operationalized in considering environmental justice (Ballet, Koffi & Pelenc, 2013) and the concept of 'sustainability' itself (Sen, 2013), and in terms of mobility it can be understood as involving a refocusing on 'what mobility is for', with the mode of travel used being less relevant than achieving functionings.

Localization and its implications

In terms of planning, an autonomobility regime would encompass known sustainability strategies, including densification, decentralization and hub-centered urbanism (Banister, 2005; Wheeler & Beatley, 2014). This would also require radical localization of the productive economy.

In a situation of eco-localization, the right to stay put in one's chosen spatial environment, raises environmental justice rights to safe and healthy living space. From this follows the right to "be protected from uncontrolled investment and growth, pollution, land grabbing, speculation, disinvestment, and decay and abandonment" (Anguelovski, 2014, p. 60). This would likely require the de-privatization and de-commodification of land, housing and associated infrastructures.

Decoupling from automobility's spatially distanced 'life chances' and social participation also involves tackling the totalizing compulsion of wage labor relations to radically reduce multiple requirements to travel. The exogenous need to commute long distances or move for employment reasons would have to dissolve and residence choices be decoupled from job opportunities, profit-centered housing markets or other economic compulsions. The unconditional basic living income is already an initial operationalization of this degrowth agenda (D'Alisa, Demaria & Kallis, 2014) to prioritize wellbeing over GDP, one which would also reduce the spatiotemporal pressures that make car use so essential (Cass & Faulconbridge, 2016).

Given that some extended chains of socioeconomic interrelations, interdependences and functional differentiations will require sustaining, some mobility will remain essential and socially desirable. This will be the case even under a material and structural foundation of economic degrowth. Environmentally, motorized transport would nevertheless have to be reduced to an essential, sustainable minimum.[4] This will likely make *individualized* transport (beyond active modes of walking and cycling) at first economically exclusive and beyond the means of most, and thereby inimical to distributional justice equity. Forms of low-impact, fast and convenient individualized transport such as electric velomobiles (more energy efficient than electric cars) might emerge as a transition mobility mode (De Decker, 2010). At larger scales, speed and environmental impacts being linked, *slow forms of mobility* and transport would become normal (Schor, 2014), which would help reverse the economically induced acceleration of everyday lives (Rosa, 2003) and counter potential social injustice and/or mobility-warlordism.

Collective mobility

The decoupling of mobility and the dominant economic paradigm, and replacing the liberal myth of individual freedoms as expressed through automobility while still promoting and extending autonomy and equity in relational networks, are far from simple! The end of the 'economic growth of production' regime, however, could open new options for more sustainable collectivized mobility regimes. A collective and socialized autonomobility system would move beyond public transport *provision* to encompass demand-responsive transport organized organically from the bottom-up, involving norms of expected sharing and rights of piggybacking on those transport mobilities that would remain as social necessities. These ideas will be further explored in the section on "seeds of the future" below.

Glimpses and visions of autonomobility

In summary, the autonomobility system is admittedly utopian and idealized. It is an egalitarian model encompassing the social relations and norms required to organize society around equality, justice and freedom under conditions of scarcity of resources but abundance of individual time. There are historical and fictional accounts of mobility systems resembling our ideas. Documented *historical* precursors include the transitory arrangements of Civil War Barcelona, where *all* social practices and norms, including mobility, were reorganized on radically communalized and egalitarian principles. Private transport was largely confiscated for the war effort, and the existing public transport system was transformed such that in days, "700 tramcars, instead of the usual 600 ... were operating. ... With the profit motive gone, safety became more important and the number of accidents was reduced. Fares were lowered and services improved" (Conlon & MacSimóin, 1986, p. 6).

An elaborated *fictional* account of autonomobility is *The Dispossessed* by Ursula Le Guin (1978), in which the moon Annares has been settled by colonists with an ideology of egalitarianism adapted to their hostile environment of scarcity. Longer-distance travel is either between communalized accommodation in settlements, mostly for variety-seeking in pursuing an occupation, by free-riding on trains transporting goods or else on electric trams within urban areas. The society's founder, Odo, has a vision, paraphrased, that communities' size be limited by their capacity to produce food and power from the immediate area and that communication and transport networks connect all communities for easy administration and the mobility of people and ideas, to include them in interchange (Le Guin, 1978, p. 85). We are not the first to highlight the inspiration the novel offers for sustainability theory and practical aspects of eco-localization, carbon footprinting, radical decentralization and so forth (Armstrong, 2007), but it also highlights the requisite shift in values. This involves moving from a pursuit of individualistic freedom to a freedom from ownership and competition, leading to a desire to do the necessary and a joy in doing it.

Prefigurative seeds of the future autonomobility

To anchor our imaginary of autonomobility in the present, we wish to highlight some promising seeds of practice and policy as examples of prefiguration (Graeber, 2004), where actions in the present build a new society in the shell of the old. We focus on active, collectivized and socialized mobility modalities, including piggybacking and 'free-riding', drawing on historical as well as global examples to identify that aspects of autonomobility have always existed.

Active and transitional mobilities

In reconfigured autonomobility landscapes, distances to economic and social goods and services should be reduced such that active modes including walking and cycling are sufficient to access them. Schools in the United Kingdom have trialed 'walking buses'; collective pedestrian commutes supervised by staff and parents. Bike-buddy schemes encourage cycle co-commuting where it is practicable, offering immediate support and sociality. Here collectivization acts to re-socialize alternative forms of mobility that have become threatened and perceived as risky through the atomization of individuals undertaking them within the restraints of automobility outlined above. Electric bikes (Behrendt, 2017) and velomobiles show signs of providing transitional modes of transport that bridge between the car and the cycle. Velomobiles can reach similar speeds, climb hills, keep people dry and carry luggage whilst also providing privacy and autonomy: "velomobiles have a speed and range ... comparable to ... electric cars ... a quarter of the existent [US] wind turbines would suffice to power as many electric velomobiles as there are people" (De Decker, 2012). Autonomobility focuses mobility policy on the materials and meanings of active mobility as social practices (Watson, 2013) – an area in which mobilities studies are already advanced. For example, the 'cargo function' of cars (Mattioli & Anable, 2015) must be supplanted by improving the carrying capacities of people and cycles, shifting clothing and punctuality norms and adjusting street architecture to fit an actively mobile population.

Collectivized and socialized mobilities

Our utopian examples of autonomobility systems relied heavily on the pre-existence of collectivization and the understanding that socialized mobility is part of an all-encompassing social drive.[5] Orwell (1952) highlights how individual movement in the Spanish Civil War piggybacked on essential transports of goods, munitions or fighters and, therefore, was considered a legitimate use of existing mobilities as public goods. Such free-riding is delegitimized under neoliberal capitalism, but is valorized in autonomobility. It thus fits mobility to transport-as-a-social-good distributions, rather than *vice versa* under distributive models. Hitchhiking is among the most well-known and widespread modes along with fare-dodging and freight-train-riding, the previous preserve of hobos and tramps (Cresswell, 1999). Such practices could be relegitimized and valorized not only through provision of suitable facilities (signposted and/or weatherproof 'auto-stops' as provided in some European countries), but also through legislating norms of lift-giving: 'single-occupancy vehicles' are anathema to autonomobile society. The preferential treatment of professional driving 'red plates' and 'tachograph' hitchers in the United Kingdom, and soldiers, carpenters and tradespeople in other countries, demonstrates that norms of rewarding contributions to society with a right to expect help with travel already seem to be latent in society. Hitchhiking in the United Kingdom as well as in most other Western societies was 'killed' primarily by a combination of circulating discourses of danger, but crucially also profit-seeking: haulage companies avoiding insurance costs for competition. Such factors could be reversed. Beyond formal lift or car-sharing and carpooling initiatives (Baptista, Melo & Rolim, 2014), many examples of more sustainable bottom-up and demand-responsive shared mobilities exist. Employers, religions, political groups and campaigners, sports fans, schools and state authorities all organize collective transport in minibuses, vans, coaches, taxis or trains. This is the normal experience of the commute in many parts of the world, particularly for flexible labor, but also in the semiformal systems of Hong Kong (Cullinane & Cullinane, 2003). The cost per individual of such systems can be cheaper than public transport, and they can be more time efficient. Demand-responsive public transport is a mode used by UK local authorities to plug the gaps in provision created by profit-driven 'public' transport (Cass, Shove & Urry, 2003, 2005), acknowledging mobility for social inclusion as a right.

In addition, collectivization of service industries via incentivization, tax breaks and so forth could move away from a dominant model of 'one-person, one-van' businesses to something more resembling a 'dial-a-service' where locality and proximity are prioritized from a register of available skills. Structural and economic changes may force these developments exogenously. The rise of the 'platform' economy has already demonstrated something of the sort – but in pursuit of corporate profits.[6] Such collectives might either represent or develop from a transition towards local sustainability or intentional eco-localization, in which such skills would be more valued and also more diffused through communities. They would require a transition from individualistic competition to collectivization, coordination and trust. The operation of fleets of vehicles for reasons of efficiency and economies of scale, with mechanics employed to maintain the fleet and ensure that variegated and flexible mobility needs can be satisfied at short notice, would be essential to a transition to an autonomobility system. Also required are changes to insurance and ownership restrictions and to the individualization of responsibility for (not only, but importantly) work-related travel. Much of this could be initially addressed by encouraging and permitting communal ownership and maintenance of vehicles and by making trip purposes and the social role they play the subject of insurance, coordination and facilitation or remuneration, rather than individual vehicles.

Counter-forces to autonomobility

Of course, there are manifold barriers to our vision of autonomobility futures. Whole literatures are devoted to identifying the conditions under which 'transitions' to more sustainable socio-technical systems take place (Geels, 2005). These usually rely on exogenous changes to 'landscape' factors to enable niche developments to displace incumbent 'regimes', and there are arguably incipient signs of such changes: the environmental crises, but also the decline of coal, disinvestment from fossil fuels, post-carbon manifestos from fossil fuel companies, investment in electric vehicles and energy storage. Many of these were unthinkable a matter of years ago. Without doubt, fossil fuel industries remain as major incumbents, but are investing in renewables. In transition, car manufacturers could turn to producing lightweight velomobiles for individual travel. The 'coevolution' of needs along with technologies that not only satisfy but create them has produced the inefficiency of the heavy, smart automobile; no more fuel efficient than its 1950s' equivalent despite huge increases in engine efficiency (Shove, 2017). Electric vechicles would pursue the same trajectory, but autonomobility requires that such needs and expectations descend with the material demands of vehicular solutions. In such ways, unsustainable path dependencies can be escaped. At root, the main roadblock is that all social systems are currently organized to satisfy shareholders, not mobility requirements or other understandings of social need, including those of a distributional model of mobility justice.

Conclusions

We outlined an imaginary of a future mobility system that would reconcile the severe demands of environmental, social and economic sustainability. Such a system would involve much beyond mobility or transport *per se*, including a rethinking of growth, production, livelihoods, income/welfare, land use, urbanism and much more, *and* the ethical, conceptual and normative changes that would support such a social system. This underlines that mobility is a derived need, and transport merely the mode of its satisfaction. Mobility satisfies needs, but is not a good in itself. For this reason we think that conceptions of mobility justice that focus on redistributing mobility (usually *qua* transport) on abstracted geographical, gender, racial or class bases misses the point of the 'good(s)' being addressed. To (re)distribute access to goods, services and opportunities that have been distantiated by automobile capitalism, whether via addressing human mobility or sites' accessibility, is still to link mobility with a neoliberal nexus of freedom, choice and happiness through material acquisition. Rather, we suggest that positive and negative conceptions of liberty (*from* compulsions, *to* pursue ends) should be enhanced along with a capabilities approach to individual flourishing and that society should be reconfigured to maximize these possibilities. The increasing restrictions implied by a movement away from (particularly individualized) motorized transport will necessitate rethinking social values attached to ownership, occupations and livelihoods. Collectivity and sociality in mobility, as in other spheres of life, will be mainstreamed. We have demonstrated that there are examples of *quotidian* collective and/or socialized mobility practices sidelined under automobility that could re-emerge. Mobility in the future will increasingly take place in a 'state of exception', and the rationalities behind these practices could increasingly overcome the individualizing, competitive drivers we have identified in automobility. *The Dispossessed* (Le Guin, 1978), where all practical aspects of the organization of society are tied to an ethical balancing of justice and freedom, under conditions of scarcity, is one available parable for the post-oil future.

Notes

1 Although automobile spatial orderings are more pronounced in the United States and Australia than in most parts of Europe, spatial and urban planning strategies and housing developments worldwide reflect the automobile imperative; that is, the car is seen as the normal mode of transport even as it passes its zenith of acceptability.
2 Jobseekers in the United Kingdom are expected to consider jobs that require up to three hours' travel a day to access.
3 'Cobalt, the heart of darkness in the shiny electric vehicle story'. *The Globe and Mail*. Retrieved from www.theglobeandmail.com/globe-investor/investment-ideas/cobalt-the-heart-of-darkness-in-the-shiny-electric-vehicle-story/article37109931/.

4 The debate as to what constitutes 'essential' mobility is intrinsically political and likely to become increasingly contested.
5 Conlon & MacSimóin (1986) note that 6,500 of the 7,000 tram workers in Barcelona were members of the anarcho-communist union CNT, and they cite Bolloten's (1961) observation that amongst 3,000 collectivized enterprises were railways, trains, buses, taxis and shipping, as well as car factories.
6 In addition, the spreading of local repair workshops and visions of open-source technologies as well as the shift to long-lasting, high-quality products will contribute to reducing the mobility of goods.

References

Aftabuzzaman, M. & Mazloumi, E. (2011). 'Achieving sustainable urban transport mobility in post peak oil era'. *Transport Policy*, 18(5), 695–702.

Aglietta, M. (2000). *A theory of capitalist regulation: The US experience*. London: Verso.

Anguelovski, I. (2014). 'Environmental justice'. In G. D'Alisa, F. Demaria & G. Kallis (Eds.), *Degrowth: A vocabulary for a new era* (pp. 60–63). London: Routledge.

Armstrong, D. (2007). 'A review of Ursula Le Guin's sci-fi classic *The Dispossessed*: An ambiguous utopia'. Retrieved from www.mudcitypress.com/muddispossessed.html.

Ballet, J., Koffi, J. & Pelenc, J. (2013). 'Environment, justice and the capability approach'. *Ecological Economics*, 85(1), 28–34.

Banister, D. (2005). *Unsustainable transport: City transport in the new century*. Abingdon, UK: Taylor & Francis.

Baptista, P., Melo, S. & Rolim, C. (2014). 'Energy, environmental and mobility impacts of car-sharing systems: Empirical results from Lisbon, Portugal'. *Procedia-Social & Behavioral Sciences*, 111, 28–37.

Behrendt, F. (2017). 'Why cycling matters for electric mobility: Towards diverse, active and sustainable e-mobilities'. *Mobilities*, 13(1), 1–17.

Berlin, I. (1969). 'Two concepts of liberty'. In I. Berlin (Ed.), *Four essays on liberty* (pp. 121–154). Oxford: Clarendon Press.

Better Transport. (2017). 'Environmental quality, climate change and transport innovation'. Retrieved from www.bettertransport.org.uk/sites/default/files/pdfs/Tracks-Carbon-Reduction-Report-2017.pdf.

Bolloten, B. (1961). *The grand camouflage: The communist conspiracy in the Spanish Civil War*. New York: Praeger.

Bonin, H., Eichhorst, W., Florman, C., Hansen, M., Skiold, L., Stuhler, J., ... Zimmerman, K. (2008). *Geographic mobility in the European Union: Optimizing its economic and social benefits*. IZA Research Report 19. Bonn: Institute for the Study of Labor.

Cass, N. & Faulconbridge, J. (2016). 'Commuting practices: New insights into modal shift from theories of social practice'. *Transport Policy*, 45(1), 1–14.

Cass, N., Shove, E. & Urry, J. (2005). 'Social exclusion, mobility and access'. *The Sociological Review*, 53(3), 539–555.

Cass, N., Shove, E. & Urry, J. (2003). *Changing infrastructures, measuring socio-spatial inclusion/exclusion*. Report for the DfT. Lancaster: Lancaster University. Retrieved from http://wp.lancs.ac.uk/elizabeth-shove/files/2014/05/chimereport.pdf.

Castells, M. (2002). 'The urban ideology'. In I. Susser (Ed.), *The Castells reader on cities and social theory* (pp. 45–70). Oxford: Blackwell.

Conlon, E. & MacSimóin, A. (1986). *The Spanish civil war: Anarchism in action.* Dublin: Workers Soldarity Movement.

Cresswell, T. (1999). 'Embodiment, power and the politics of mobility: The case of female tramps and hobos'. *Transactions of the Institute of British Geographers,* 24 (2), 175–192.

Cullinane, S. & Cullinane, K. (2003). 'Car dependence in a public transport dominated city: Evidence from Hong Kong'. *Transportation Research Part D: Transport & Environment,* 8(2), 129–138.

D'Alisa, G., Demaria, F. & Kallis, G. (2014). *Degrowth: A vocabulary for a new era.* London: Routledge.

De Decker, K. (2012). 'Electric velomobiles: As fast and comfortable as automobiles, but 80 times more efficient'. In *Low-Tech Magazine.* Retrieved from www.low techmagazine.com/2012/10/electric-velomobiles.html.

De Decker, K. (2010). 'The velomobile: high-tech bike or low-tech car?' *Low-Tech Magazine.* Retrieved from www.lowtechmagazine.com/2010/09/the-velomobile-high-tech-bike-or-low-tech-car.html#more.

Delbosc, A. & Currie, G. (2011). 'The spatial context of transport disadvantage, social exclusion and well-being'. *Journal of Transport Geography,* 19(6), 1130–1137.

Dennis, K. & Urry, J. (2009). *After the car.* Cambridge: Polity Press.

Geels, F. W. (2005). 'The dynamics of transitions in socio-technical systems: A multi-level analysis of the transition pathway from horse-drawn carriages to automobiles (1860–1930)'. *Technology Analysis & Strategic Management,* 17(4), 445–476.

Golub, A. & Martens, K. (2014). 'Using principles of justice to assess the modal equity of regional transportation plans'. *Journal of Transport Geography,* 41, 10–20.

Goodwin, P. & Van Dender, K. (2013). '"Peak car" – themes and issues'. *Transport Reviews,* 33(3), 243–254.

Gordon, U. (2008). *Anarchy alive! Anti-authoritarian politics from practice to theory.* London: Pluto Press.

Graeber, D. (2004). *Fragments of an anarchist anthropology.* Chicago, IL: Prickly Paradigm Press.

Graham, S. & Marvin, S. (2001). *Splintering urbanism: Networked infrastructures, technological mobilities and the urban condition.* London: Routledge.

Harvey, D. (2001). *Spaces of capital: Towards a critical geography.* Abingdon, UK: Routledge.

Harvey, D. (2000). *Spaces of hope.* Edinburgh: Edinburgh University Press.

Havard, S., Reich, B. J., Bean, K. & Chaix, B. (2011). 'Social inequalities in residential exposure to road traffic noise: An environmental justice analysis based on the RECORD Cohort Study'. *Occupational & Environmental Medicine,* 68(5), 366–374.

Huxley, M. (2006). 'Spatial rationalities: Order, environment, evolution and government'. *Social & Cultural Geography,* 7(5), 771–787.

Le Guin, U. K. (1978). *The dispossessed: An ambiguous utopia.* London: Panther Granada.

Lucas, K. (2004). *Running on empty: Transport, social exclusion and environmental justice.* Bristol, UK: Policy Press.

Manderscheid, K. (2018). 'From the automobile to the driven subject?' *Transfers,* 8(1), 24–43.

Manderscheid, K. (2014). 'The movement problem, the car and future mobility regimes: Automobility as dispositif and mode of regulation'. *Mobilities,* 9(4), 604–626.

Martens, K. (2016). *Transport justice: Designing fair transportation systems.* Abingdon, UK: Routledge.

Martens, K. (2012). 'Justice in transport as justice in accessibility: Applying Walzer's "spheres of justice" to the transport sector'. *Transportation*, 39(6), 1035–1053.

Mattioli, G. & Anable, J. (2015). *Carrying capacity: The cargo function of the cart.* Lancaster, UK: Demand Centre. Retrieved from www.demand.ac.uk/wp-content/up loads/2016/10/DEMAND-insight-3.pdf.

Mattioli, G., Anable, J. & Vrotsou, K. (2016). 'Car dependent practices: Findings from a sequence pattern mining study of UK time use data'. *Transportation Research Part A: Policy & Practice*, 89(Supp. C), 56–72.

Mitchell, G. & Dorling, D. (2003). 'An environmental justice analysis of British air quality'. *Environment & Planning A*, 35(5), 909–929.

Mouffe, C. (2011). *On the political.* Abingdon, UK: Routledge.

Mullen, C. & Marsden, G. (2018). 'The car as a safety-net: Narrative accounts of the role of energy intensive transport in conditions of housing and employment uncertainty'. In A. Hu, R. Day & G. Walker (Eds.), *Demanding energy: Space, time and change* (pp. 145–164). Cham: Springer International Publishing.

Newman, D. (2017). 'Automobiles and socioeconomic sustainability: Do we need a mobility bill of rights?' *Transfers*, 7(2), 100–106.

Newport, D., Chesnes, T. & Lindner, A. (2003). 'The "environmental sustainability" problem: Ensuring that sustainability stands on three legs'. *International Journal of Sustainability in Higher Education*, 4(4), 357–363.

North, P. (2010). 'Eco-localization as a progressive response to peak oil and climate change – a sympathetic critique'. *Geoforum*, 41(4), 585–594.

Nussbaum, M. (2001). *Women and human development: The capabilities approach.* Cambridge: Cambridge University Press.

OECD. (2000, October). 'Environmentally sustainable transport: Futures, strategies and best practices'. International EST Conference, Vienna, Austria.

Orwell, G. (1952). *Homage to Catalonia.* Boston, MA: Houghton Mifflin Harcourt.

Paterson, M. (2007). *Automobile politics: Ecology and cultural political economy.* Cambridge: Cambridge University Press.

Pereira, R., Schwanen, T. & Banister, D. (2017). 'Distributive justice and equity in transportation'. *Transport Reviews*, 37(2), 170–191.

Peters, P. (2006). *Time, innovation and mobilities: Travel in technological cultures.* Abingdon, UK: Routledge.

Rajan, S. (2007). 'Automobility, liberalism and the ethics of driving'. *Environmental Ethics*, 29(1), 77–90.

Rammler, S. (2008). 'The *Wahlverwandtschaft* of modernity and mobility'. In W. Canzler, V. Kaufmann & S. Kesselring (Eds.), *Tracing mobilities: Towards a cos-mobilitan perspective* (pp. 57–76). Aldershot, UK: Ashgate.

Rosa, H. (2003). 'Social acceleration: Ethical and political consequences of a desynchronized high-speed society'. *Constellations*, 10(1), 3–33.

Rose, N. (1999). *Powers of freedom: Reframing political thought.* Cambridge: Cambridge University Press.

Rose, N. (1996). 'The death of the social? Re-figuring the territory of government'. *International Journal of Human Resource Management*, 25(3), 327–356.

Schneidewind, U. (2015). 'Transformative Wissenschaft: Motor für gute Wissenschaft und lebendige Demokratie'. *GAIA-Ecological Perspectives for Science & Society*, 24 (2), 88–91.

Schor, J. (2014). 'Work sharing'. In G. D'Alisa, F. Demaria & G. Kallis (Eds.), *Degrowth: A vocabulary for a new era* (pp. 195–198). London: Routledge.

Sen, A. (2013). 'The ends and means of sustainability'. *Journal of Human Development & Capabilities,* 14(1), 6–20.

Shove, E. (2017). 'What is wrong with energy efficiency?' *Building Research & Information.* doi:10.1080/09613218.2017.1361746.

Simcock, N. & Mullen, C. (2016). 'Energy demand for everyday mobility and domestic life: Exploring the justice implications'. *Energy Research & Social Science,* 18, 1–6.

Ureta, S. (2008). 'To move or not to move? Social exclusion, accessibility and daily mobility among the low-income population in Santiago, Chile'. *Mobilities,* 3(2), 269–289.

Urry, J. (2013). *Societies beyond oil: Oil dregs and social futures.* London: Zed Books.

Urry, J. (2007). *Mobilities.* Cambridge: Polity Press.

Urry, J. (2004). 'The "system" of automobility'. *Theory, Culture & Society,* 21(4–5), 25–39.

Urry, J. (2002). 'Mobility and proximity'. *Sociology,* 36(2), 255–274.

Wadud, Z., MacKenzie, D. & Leiby, P. (2016). 'Help or hindrance? The travel, energy and carbon impacts of highly automated vehicles'. *Transportation Research Part A: Policy & Practice,* 86, 1–18.

Watson, M. (2013). 'Building future systems of velomobility'. In E. Shove & N. Spurling (Eds.), *Sustainable practices: Social theory and climate change* (pp. 117–131). Oxford: Routledge.

Wheeler, S. M. & Beatley, T. (2014). *Sustainable urban development reader.* London: Routledge.

Zhang, K. & Batterman, S. (2013). 'Air pollution and health risks due to vehicle traffic'. *Science of the Total Environment,* 450, 307–316.

8 Dark design

Mobility injustice materialized

Ole B. Jensen[1]

Introduction

Foucault wrote,

> a whole history remains to be written of spaces – which would at the same time be the history of powers (both these terms in the plural) – from the great strategies of geo-politics to the little tactics of the habitat, institutional architecture from the classroom to the design of hospitals, passing via economic and political installations.
>
> (1980, p. 149)

As Foucault suggests, design ideas and practices are based on normative principles and values, which may be more or less explicit. In this chapter, I explore the dynamics of 'dark design' within the area of mobility, showing how design ideas and interventions deliberately produce social exclusion by regulating movement in particular ways. The key focus is exploring examples of design where power (primarily manifested as social exclusion) resides in complex (and sometimes subtle) relationships to the materialization of artifacts, systems and design. The notion of dark design enables us to question taken-for-granted assumptions about spaces and materials as 'simply there'. Instead, I show how social exclusion and questions of justice are embedded in design choices and solutions. There are many socially inclusive forms of design; however, here light is thrown on exclusionary design practices and the darkness of design.

The chapter connects thinking about design codes, codes of conduct, city plans, technologies, artifacts, objects and materialities (e.g., Anderson & Wylie, 2009; Bennett, 2010; Ingold, 2014; Jensen, 2016; Jensen & Lanng, 2017; Latour & Yaneva, 2008; Richardson & Jensen, 2008) to notions of mobility justice (Cook & Butz, 2016; Sheller, 2012, 2014). Empirically, it spans from the extreme of the Holocaust to the dark design of mundane city spaces such as squares and public spaces, where the confinement and exclusion of 'unwanted subjects' is achieved and mediated through 'bum-proof' urban furniture, fences and spikes that move them along and keep them out of particular places. The imprint of social interests is detectable in all aspects of design from the grand designs of states (Scott, 1998) to minute interventions on the sidewalk that control human action (Lofland, 1998).

After developing the notions of dark design and mobility justice, I discuss 'new materialities' from the perspective of the 'mobilities turn'. Then I analyze cases of dark design reaching from the Holocaust to the 'warfare' against homeless people in urban spaces. The chapter ends with concluding reflections about the intersection of dark design and mobility, and issues for future research.

Framing dark design: between (in)justice and materialities

The first task is to address the term 'dark'. Social scientists frequently employ the metaphor 'dark' when discussing issues of power and social exclusion (e.g., Flyvbjerg, 1996; Foucault, 1975). This tradition draws threads back to Machiavelli and the medieval philosophy of power. Darkness was connected to the 'uncanny' and hidden powers at work. Dark design also implies a normative positioning and standpoint. I cannot speak of injustice and exclusion without articulating a yardstick for the discussion. A critical analysis of dark design thus relates how it violates and threatens values like 'human flourishing' (Freidman, 2002), 'multicultural tolerance' (Sandercock, 2003), 'interspecies co-existence' (Haraway, 2016), 'rights to the city' (Lefebvre, 1996) and 'spatial justice' (Soja, 2000). The normative compass for speaking of dark design is grounded in ideas of human flourishing and critical, emancipatory practices. More practically, I am inspired by the project *Design like you give a damn* (Architecture for Humanity, 2006), the so-called *Critical Engineering Manifesto* (The Critical Engineering Working Group, 2011) and Sayer's (2011) call for social science to better understand "why things matter to people".

Mobility justice

The first leg of our theoretical framework is the notion of 'mobility justice'. One of the key thinkers in the field, Mimi Sheller (2014), defines the concept in the following manner:

> Mobility justice, for me, is a way of thinking about differential mobilities, and thinking about the ways in which people's mobilities are interrelated, and that we have different capabilities for mobility and different potentials for mobility. It's not to say that being mobile is always good, or that it equates with freedom, because sometimes there are coerced forms of mobility, and sometimes staying still is important – being able to remain in a place. So mobility justice is a way to talk about those different relations around mobility, and that is a way to highlight the power differentials that come into play in any form of mobility, and the different affordances that different people are able to make use of, or appropriate, in becoming mobile or not.

Mobility justice is a complex matter. Mobilities may need to be restricted for the benefit of the commons (as when restrictions on cars are implemented in overburdened city centers) and, therefore, cannot necessarily be equated with the unfettered freedom to move and 'right to roam'. And differential or striated mobilities are contextual. In other words, we need to understand the contextual complexity of specific mobility systems and practices in order to see if we are dealing with violations of mobility justice. For example, Martens (2012, p. 3), who theorizes transportation justice using Walzer's notion of distributive justice, claims that "there can be no single criterion in virtue of which all goods are to be made available to members of society".

The right to roam or individual freedom of movement is frequently connected to issues of justice and power (Ernste, Martens & Schapendonk, 2012; Martens, 2012; Sheller, 2012, 2017). Kaufmann (2002) illustrates that the capacity to move (motility) is an important social resource that is connected to the exercise of power (as well as to processes of social exclusion). The right to roam is the material manifestation of human autonomy, the restriction of which is often understood as an act of coercion. However, as much as individual free movement may be felt as a right, a relational understanding of mobility justice clearly illustrates that this is not so simple. At times, someone's mobility is relationally interdependent with others' immobility – as, for example, when roads and infrastructures are built between cities or neighborhoods with so-called 'barrier effects', enhancing the mobility of drivers while blocking previously rich spatial interactions between neighbors. Danish philosopher K. E. Løgstrup (1991, p. 25) argues that we never encounter other humans without "holding something of their lives in our hands". In transport, as in other spheres of life, we are thus relationally interdependent. This means we are facing what Løgstrup terms an 'ethical demand' to act responsibly. The relational interdependency of (mobile) social actors is also connected to complex re-distributional questions as the systems and infrastructures affording mobilities may have inbuilt effects related to the inequitable experience of social goods and bads. Moreover, if we consider environmental dimensions, the future consequences of our mobility choices and policies become even more complex. In sum, we should have to balance individual freedom of movement with repercussions for the contemporary and future commons before we are able to judge the political and ethical effects of choices and interventions.

Philosopher Iris Marion Young has criticized the notion that social justice is limited to the sphere of distribution. She argues for adding the notion of domination to get a fuller understanding of the question of justice. Drawing on this approach, Cook and Butz (2016, p. 403) propose that the societal goods of infrastructure provision and individual mobility may be seen as more differentiated:

> Young's theory of domination ... leads us to conceptualize mobility justice not only as the equitable distribution of motility throughout a social

system, but also the elimination of domination in the political field of mobility, or in other words as just institutional actions and decision-making processes about mobility issues that promote just mobility outcomes.

The notion of domination enables an analysis that moves beyond the equitable distribution of motility towards issues of suppression, exclusion, marginalization and identity denial. Based on Young's understanding of domination, it can be seen how domination enables a systematic description of "forces within institutional contexts that preclude people from participating in procedures that establish possibilities and conditions for action" (Cook & Butz, 2016, p. 403).

The emerging research position on mobility justice thus articulates a sensitivity to differential and relational mobilities, access to mobility and the right to move and distributions of capacities to move as well as to questions of domination and social exclusion. Before exploring empirical examples of dark design relevant to some of these aspects of mobility justice, I look into the second part of the theoretical frame that enables us to understand more fully how such injustices are instituted and materialized through design.

Mobilities, materialities and design

One of the key insights of the 'turn to mobilities' (e.g., Cresswell, 2006; Urry, 2007) is that the systems and designs affording mobilities are much more than simply infrastructures of movement. Ernste, Martens and Schapendonk (2012, p. 509) make this point with reference to mainstream transport research that sees mobility as a 'derived demand', in contrast to mobilities thinking that understands it as a meaningful social practice in and of itself. Another central mobilities turn claim is that we need to focus not just on that which moves, but also on the nexus of mobility and immobility. In Urry's (2003, p. 126) words, "it is the dialectic of mobility/moorings that produces social complexity". This dialectic is crucial to the discussion that follows, since much of the dark design of mobilities is as much about blocking and restricting movement as it is about channeling and directing it through designed systems and infrastructures. Systems of mobilities are sites of interaction, places of public life and arenas of communal life. Therefore, mobility systems are sites of cultural and identity formation as well as instrumental places of physical movement. However, if these sites of mobilities serve purposes other than simply moving people, then access to them will have profound consequences for citizenship and quality of life. Hence, issues of social stratification, exclusion and injustice connect with dimensions of mobilities. The ways in which mobility infrastructures are materially configured are important clues to understanding how injustice and social exclusion are instituted and materialized through design decisions and interventions. Ernste, Martens and Schapendonk (2012, p. 510) also see the connection between mobility justice and design when they argue that "[t]ransport mobility research could pay more attention to the social design of mobility and the way it shapes patterns of movement and non-movement" and that "different designs create different mobilities".

Mobility (in)justice materializes through design by its ability to configure who can and cannot move (as well as, in more extreme cases, who is forced to move). The level of 'darkness', or social injustice, and the directness with which it is materialized into mobility spaces and infrastructures may vary. In some instances, design is part of a "mundane politics" (Yaneva, 2017). In other words, subtle and less explicit examples of power manifest in design – what design analyst Lars Frers (2009) terms "pacification by design" – through 'silent participation' in the assemblages and networks of social exclusion. Designs may nudge people in specific directions; they may hide things or expose them. In Frers' terminology, these small acts of design contribute to the production of "localized normalities" (2009, p. 247). The orchestration and design of such localized normalities relate to the mobilities turn insight that landscapes of mobility infrastructures are 'habitats' of contemporary urbanites and, thus, 'normalized scenes' of everyday life. This means that the subtle and normalized acts of pacification by design lie at one end of a continuum of dark design.

However, mobility injustice can be produced in ways that are more explicit. To understand this, I turn to geographer Steven Flusty, who defines five spaces designed to exclude unwanted subjects and social practices:

- stealthy spaces, which are camouflaged or obscured and hidden;
- slippery spaces, which are unreachable due to contorted, protracted or missing paths of approach;
- crusty spaces, which are inaccessible due to obstructions such as walls, gates and checkpoints;
- prickly spaces, which are uncomfortable due to sloped ledges; and
- jittery spaces, which are unusable due to active surveillance.

(List from Carmona, Tiesdell, Heath & Oc, 2010, p. 156)

These exclusionary spaces are employed, for instance, in the 'war against homeless', in which crusty spaces may be mixed with both prickly and jittery spaces in very explicit expressions of dark design. Lofland (1998, p. 190) reminds us why formal acts of regulation are insufficient to modify people's behavior and why design interventions are so compelling:

> If regulation alone could achieve the purification of the public realm, we would all currently live in a world from which not only the homeless, street prostitutes, peddlers and drunkards had completely disappeared, but so had such diverse activities as panhandling and begging, loitering, rollerskating and skateboarding, singing, shouting, eating, soliciting, dancing, shilling, parading and protesting, miming, making music, politicking,

courting, urinating, swearing or cursing, fighting, gambling, spitting and game playing.

The manifestation of dark design may thus come about through complex combinations of governmental regulations and detailed design interventions. I now explore examples of dark design through the framework of two dimensions of mobility justice and materialites in order to enable a deeper understanding of how mobility injustice is materialized.

Lessons in dark design

Not all dark designs are equally hurtful or important in relation to mobility injustice; rather, they exist on a continuum. Therefore, I differentiate between an extreme case of dark design (the Holocaust) and mundane examples; in both instances power is exercised to exclude unwanted subjects.

Holocaust – the ultimate illustration of dark design

The ultimate case or 'Ground Zero' of dark design is the death camps of the Holocaust.[2] As Bauman (1994) argues, there are extraordinary and 'rational' engineering logics underlying this most dehumanizing of all historical events. The Holocaust is thus an extreme example of dark design in which the ideologies of in-humanism were deeply embedded into the logistic protocols of human movement and the engineering features of extinction technologies. In the context of this short chapter, I cannot possibly devote sufficient attention to the magnitude of the Holocaust. However, in the context of dark design, its complex orchestration and logistics of human movement seem to be defining and pivoting themes. Historians have demonstrated that the turning point in the systematic process of mass murder came when the Nazis abandoned mobile death units and began building stationary camps dedicated to killing 'unwanted subjects', Auschwitz being the most infamous example (Browning, 1986). This new mode of genocide involved moving large numbers of people by busses and later trains to stationary camps, an operation that succeeded through a unique combination of engineering skills, infrastructure development, logistics and human mobility. From the archives, it can be seen how train timetables were meticulously coordinated with the capacities of gas chambers, and how the reordering of humans from smaller extinction units into large factory-like systems was a powerful manifestation of efficiency and rationalization.

For example, the so-called 'Operation Reinhardt' in 1941 entailed a plan in which mobilities and design played key roles in systematically terminating Jews (Arad, 1987; Browning, 1986). The operation involved a huge mass movement of people, and it should be seen as the dark design solution to the so-called 'Jewish problem', which Hitler said, "really was a spatial question" (Browning, 1986, p. 509). Reducing the Jewish problem to a "spatial question" led to a 'solution' dependent on logistics and mobility as a particular way of addressing the spatiality of the Jews. As part of the huge logistical operation of transporting live Jewish bodies to sites of annihilation, the selection of strategic and logistically central nodes in the Third Reich rail network minimized transportation distances. Viewed through the lens of cold rationality and the "banality of evil" (Arendt, 1963), Bauman (1994) argues that the Holocaust entailed a substitution of moral responsibility with technical performativity. This operation of human displacement and annihilation illustrates a very clear relationship between human mobilities and the materialization of injustice through dark design.

There are several touchpoints among the dark design of the Holocaust, materialities and mobility justice. One obvious dimension is the way in which the mobile orchestration of human bodies to the death camps is an expression of the infringement of the right to immobility and human autonomy. Not only are we facing issues of human suppression and identity denial, we are facing well-orchestrated programs of forced movement to genocidal ends. As we move closer to the design dimension we see how the coerced 'herding' of the masses of bodies required rather strict mass-transportation systems provided with armed guards, as well as barbed wire and mines, to support the 'channeling of bodies' into the system of movement. Many of the micro-design techniques illustrated by Flusty were operative in the process of Holocaust (e.g., crusty spaces of fenced-in areas and the jittery spaces of surveillance). These spatial and material manifestations of dark design were found in the collection camps, in the strictly policed transportation systems and in the terminal stations of the death camps.

The urban wars against homeless people

The next example of dark design is less extreme than the Holocaust, but still involves a very serious set of socially exclusionary practices and design interventions. In cities across the world, a 'war' is being fought among property owners, developers, city governments, police and homeless people as relationally interdependent 'combatants'. The tools and techniques of "urban militarization" (Graham, 2010) and social exclusion range from gated communities and environments sorted and controlled by software to bum-proof benches and spikes in publicly accessible doorways. They bear witness to the increasingly sophisticated control, sanitization and securitization of urban space. A newly homeless person provides some sense of how it feels to be subject to these materializations of social exclusion through design:

> From ubiquitous protrusions on window ledges to bus-shelter seats that pivot forward, from water sprinklers and loud muzak to hard tubular rests, from metal park benches with solid dividers to forests of pointed cement bollards under bridges, urban spaces are aggressively rejecting soft, human bodies. We see these measures all the time within our urban

environments, whether in London or Tokyo, but we fail to process their true intent. I hardly noticed them before I became homeless in 2009. An economic crisis, a death in the family, a sudden breakup and an even more sudden breakdown were all it took to go from a six-figure income to sleeping rough in the space of a year. It was only then that I started scanning my surroundings with the distinct purpose of finding shelter, and the city's barbed cruelty became clear.

(Andreau, 2015)

The statement certainly testifies to the "city's barbed cruelty", but it also identifies the ways in which the spaces (and thus the design interventions) are "aggressively rejecting soft, human bodies" (Andreau, 2015). This is the pivotal argument and the ultimate test of the darkness of design, and precisely why we need a vocabulary of materialities to support our understanding of socially exclusionary practices. Rather than being questions of procedural justice or distribution, the hurtful ways in which the spikes and barbed wires connect to, rub against and even at times penetrate soft human bodies are at the center of design justice. As design interventions work on human bodies, we start to understand how the embedded rationalities of design interventions in a very real and material sense create social exclusion and injustice (clearly this was also the case in my account of the Holocaust). These micro-designs and insertions of hurtful artifacts and objects are vivid examples of Flusty's crusty, prickly and jittery spaces, mentioned above.

The account from this homeless person resonates with the way Mike Davis (1992, p. 232) describes how Fortress LA became the embodiment of "defensible" architecture (Newman, 1996) that destroyed public space and produced what Davis termed "sadistic street environments" (1992, p. 232). The leaning and curved benches that prevent bodies from resting and sleeping combined with the timed watering of grass lawns in public parks are design components of this urban warfare against homeless people. These technologies and designs are deliberate and illustrate the particular rationalities performing through the materialities of urban furniture and urban spaces. In contemporary cities, urban furniture is designed to prevent people from lingering, and pathways lead pedestrians where designers want them to go. As we reach back to Urry's notion of the immobility/mobility dialectic and connect it to Sheller's argument about mobility differentials, we start to realize how these acts of dark design sometimes move and channel human bodies and, at other times, bring them to a standstill and halt. In the extreme case of the Holocaust this leads to human annihilation, but in terms of sadistic urban environments this regulation of movement through dark design excludes particular people from particular spaces and activities against their will.

Some of the interventions and designs directly orchestrate flows and movements by rendering benches, doorways and grass lawns uninhabitable. This is the subtle play between mobility and immobility in which the embedded rationalities of dark design may work to produce spaces for lingering and occupancy or push bodies away to elsewhere. Urban no-go areas and design blockings force movement to 'free zones', areas not yet imprinted with dark design. So while bum-proof benches and metal spikes nested into concrete are stationary and sedentary interventions and devices, they afford and enforce movement to other places, establishing an urban mosaic of 'go/no-go' areas. Places of forbidden access exist alongside places of access, creating an urban jigsaw puzzle constituted through corridors of movement/access and immobility/exclusion. Furthermore, these meticulous interventions work directly on unwanted subjects' bodies by denying them a public space of being, excluding them from this sphere of social life. Over time, such acts of citizen denial surely contribute to a general erosion of self-confidence amongst people who already are at the bottom of societies.

Dark design can be understood as hostile architecture in the sense that it works on bodies in ways just described. However, there can be reasons why planning and design at times need to impede on subjects' autonomy for the sake of the commons. So, urban planning and design walk a tightrope, as described by Ella Morton (2016):

> But where is the line drawn between hostile architecture that seeks to favor one class of people over another, and practical urban planning that aims to keep all people safe? "All urban architecture or urban design has a level of control built into it," says Petty, citing pedestrian crossings and sidewalks as features that guide the behavior of the public. "But then you've got a point where that kind of controlling becomes direct, explicit and targeted against certain groups and not others".

As mentioned, dark design, including hostile or defensive architecture, exists on a continuum. At one end are the overt design features that are obvious to anyone walking by, like spikes and fences. At the other end are the design elements in which "the hostile function is often embedded under a socially palatable function" (Morton, 2016). These are precisely the acts of mundane politics and localized normalities that work through the silent participation of artifacts and design ideas. The acts of exclusion through design interventions I have presented here are mostly connected to the themes of the 'right to roam' and 'the right to the city' (Lefebvre, 1996). However, if we think about the right to the city and the accessibility of citizens to the public realm of the city, the right to roam the city and the acceptance of all citizens as legitimate urban dwellers are compromised in many of these examples. If we consider the city as a space in which the free movement of all is a precondition to understanding oneself as a citizen, then the lack of access equally becomes an issue of social exclusion.

Concluding reflections

The examples of dark design explored here are only a few out of thousands of potential cases. They range from the extreme, as in the case of extinction camps, to the more subtle, as when certain groups are prevented access to public facilities.[3] Design, planning and architecture must be understood as always normative. Whether they qualify for the label 'dark design' is a concrete and contextual question, which in itself is a critical, normative valuation. What I wanted to address here were examples where there should be no doubt as to intentions. At the end of the day, a death camp and a doorway with spikes are intended to hurt, intimidate and exclude. The hurtfulness may be scalable, but it is there, and we need to research it!

So, here, I shall speak of the embedded rationalities of dark design as an example of the new materialities insight that things have agency, which widens our understanding of the relationship between power, materialities and design. The embedded rationalities of the designer work in complex networked relationships with materialities in place. Take, for instance, the spikes in public spaces that are meant to prevent the unwanted from occupying benches or other public places. Such materialities and artifacts are working and exercising their hurtful agency as a consequence of the rationalities of the designer. The materials have not chosen to locate themselves in these particular sites, but are meticulously and strategically inserted into the urban fabric to create socially exclusionary effects in particular situations. Put differently, we may think of these exclusionary rationalities as processual, situational and relational phenomena, which require assemblages of different (but particular) bodies, artifacts and objects in time and space. The multiple agencies at play in these matters of dark design reach from design norms and rationalities over human bodies in circulation into the material agencies of steel, concrete and barbed wire.

This chapter has illustrated the importance of exploring the complex intertwined networks of designers, intentions, ideas, values, plans, policies, controversies, human bodies, artifacts, technologies, publics, spaces and much more. In particular, the situational practices through which dark design manifests itself in concrete and material connections between humans, artifacts and systems have been highlighted through their capacities for materializing mobility injustice. Rather than ascribing the unjust and socially exclusionary effects to 'abstract systems' or 'omnipotent designers' alone, I have aimed to show how artifacts and materialities act through processes of delegation (Latour & Yaneva, 2008; Yaneva, 2009, 2017). Spikes, for example, substitute for the acts of human operators, such as private guards, police officers or janitors, who otherwise would have had to perform the socially exclusionary acts of dispelling the unwanted from the chosen sites of intervention. The capacity to make a difference in the material networks of dark design makes the chosen artifacts perform tasks in a subtle interplay between humans and nonhumans. The research into dark design must take such a situational, pragmatic and critical perspective to understand the specific coming-together of these multiple entities.

Clearly, a framework built around the two central concepts of mobility justice and materialities needs to be expanded. So too do the theoretical references for establishing such a framework. More nuance and detail need to be included in a fully fledged attempt to theorize mobilities and social exclusion by design. More subtle understandings of mobility justice are needed. Equally, we need to follow more deeply the conceptualization and theorization of materialities. In a similar vein, we also need more case studies to be conducted, partly in order to explore more cases of these explicit acts of exclusion, but also to explore some of the more subtle examples of dark design. This chapter has been a first attempt to articulate a research theme of importance in the borderland between mobilities design and mobilities justice.

Notes

1 The author wishes to thank the editors for constructive comments that improved this chapter.
2 By focusing on this extreme case of dark design, I may risk overshadowing more mundane cases that operate in our everyday lives. However, to respect Holocaust victims, including my late father, I include a discussion of it here. My father was born in Denmark in 1927 (and passed away in 2008). He was jailed in 1945 by German occupation forces for his political resistance activities, and at the age of 17 became a political prisoner in a German KZ camp. He was forced to do factory work, stack dead corpses and dig out unexploded allied bombs, from which he suffered 'post-traumatic stress disorder' (called 'KZ Syndrome' in postwar Denmark). He was granted a lifelong pension from the Danish state and retired in his mid-fifties. I dedicate this chapter to the memory of my father who lived to tell of his experiences, as well as to the millions who did not.
3 In the seminal book *Splintering Urbanism*, Graham and Marvin (2001) give an account of a city government in the United States that approached a shopping mall builder to enquire whether they wanted public transport provision, only to learn that they wanted none! Equally interesting is the discussion of Robert Moses' tunnels, which allegedly were too low to accommodate public busses mainly transporting poor people of color to the beaches of New York (Winner, 1980).

References

Anderson, B. & Wylie, J. (2009). 'On geography and materiality'. *Environment & Planning A*, 41(29), 318–335.

Andreau, A. (2015, February 19). 'Anti-homeless spikes: "Sleeping rough opened my eyes to the city's barbed cruelty"', *The Guardian*. Retrieved from www.theguardian.com/society/2015/feb/18/defensive-architecture-keeps-poverty-undeen-and-makes-us-more-hostile?CMP=share_btn_link.

Arad, Y. (1987). *Belzec, Solibor, Treblinka: The Operation Reinhard death camps*. Bloomington, IN: Indiana University Press.

Architecture for Humanity. (2006). *Design like you give a damn: Architectural responses to humanitarian crises*. London: Thames & Hudson.

Arendt, H. (1963). *Eichmann in Jerusalem: A report on the banality of evil*. New York: Viking Press.

Bauman, Z. (1994). *Modernitet og Holocaust*. København, DK: Hans Reitzels Forlag.

Bennett, J. (2010). *Vibrant matter: A political ecology of things*. Durham, NC: Duke University Press.

Browning, C. R. (1986). 'Nazi resettlement policy and the search for a solution to the Jewish question, 1939–1941'. *German Studies Review*, 9(3), 497–519.

Carmona, M., Tiesdell, S., Heath, T. & Oc, T. (2010). *Public places, urban spaces: The dimensions of urban design* (2nd Ed.). Oxford: Architectural Press.

Cook, N. & Butz, D. (2016). 'Mobility justice in the context of disaster'. *Mobilities*, 11 (3), 400–419.

Cresswell, T. (2006). *On the move: Mobility in the modern western world.* London: Routledge.

Davis, M. (1992). *City of quartz: Excavating the future in Los Angeles.* New York: Vintage Books.

Ernste, H., Martens, K. & Schapendonk, J. (2012). 'The design, experience and justice of mobility'. *Tijdschrift voor Economische en Sociale Geografie*, 103(5), 509–515.

Flyvbjerg, B. (1996). 'The dark side of planning: Rationality and "realrationalität"'. In S. J. Mandelbaum, L. Massa & R. Burchell (Eds.), *Explorations in planning theory* (pp. 383–394). New Brunswick, NJ: Rutgers University Press.

Foucault, M. (1980). *Power/knowledge: Selected interviews and other writings, 1972–1977.* London: Harvester.

Foucault, M. (1975). *Discipline and punish: The birth of the prison.* New York: Random House.

Frers, L. (2009). 'Pacification by design: An ethnography of normalization techniques'. In H. Berking, S. Frank & L. Frers (Eds.), *Negotiating urban conflicts: Interaction, space and control* (pp. 249–262). Bielefeld, DE: Transcript Verlag.

Friedman, J. (2002). *The prospects of cities.* Minneapolis, MN: University of Minnesota Press.

Graham, S. (2010). *Cities under siege: The new militarization of urbanism.* London: Verso.

Graham, S. & Marvin, S. (2001). *Splintering urbanism: Networked infrastructures, technological mobilities and the urban condition.* London: Routledge.

Haraway, D. (2016). *Staying with the trouble: Making kin in the Chthulucene.* Durham, NC: Duke University Press.

Ingold, T. (2014). 'Designing environments for life'. In K. Hastrup (Ed.), *Anthropology and nature* (pp. 233–246). London: Routledge.

Jensen, O. B. (2016). 'Of "other" materialities: Why (mobilities) design is central to the future of mobilities research'. *Mobilities*, 11(4), 587–597.

Jensen, O. B. & Lanng, D. B. (2017). *Mobilities design: Urban designs for mobile situations.* London: Routledge.

Kaufmann, V. (2002). *Re-thinking mobility: Contemporary sociology.* Aldershot, UK: Ashgate.

Latour, B. & Yaneva, A. (2008). 'Give me a gun and I will make all buildings move: An ANT's view of architecture'. In R. Geiser (Ed.), *Explorations in architecture: Teaching, design, research* (pp. 80–89). Basel, CH: Birkhäuser.

Lefebvre, H. (1996). *Writings on cities.* Oxford: Blackwell.

Lofland, L. H. (1998). *The public realm: Exploring the city's quintessential social territory.* New Brunswick, NJ: Aldine Transaction.

Løgstrup, K. E. (1991). *Den etiske fordring.* København, DK: Gyldendal.

Martens, K. (2012). 'Justice in transport as justice in access: Applying Walzer's "spheres of justice" to the transport section'. *Transportation*, 39(6), 1035–1053.

Morton, E. (2016, May 5). 'The subtle design features that make cities feel more hostile'. *Atlas Obscura*. Retrieved from www.atlasobscura.com/articles/the-subtle-de

sign-features-that-make-cities-feel-more-hostile?utm_source=facebook.com&utm_m edium=wired.

Newman, O. (1996). *Creating defensible space.* Washington, DC: US Department of Housing and Urban Development Office of Policy Development and Research.

Richardson, T. & Jensen, O. B. (2008). 'How mobility systems produce inequality: Making mobile subject types on the Bangkok sky train'. *Built Environment*, 34(2), 218–231.

Sandercock, L. (2003). *Cosmopolis II: Mongrel cities for the 21st century.* London: Continuum.

Sayer, A. (2011). *Why things matter to people: Social science, values and ethical life.* Cambridge: Cambridge University Press.

Scott, J. C. (1998). *Seeing like a state: How certain schemes to improve the human condition have failed.* New Haven, CT: Yale University Press.

Sheller, M. (2017). 'Mobilities'. In A. S. Silva (Ed.), *Dialogues on mobile communication* (pp. 51–66). Abingdon, UK: Routledge.

Sheller, M. (2014). 'Mobility justice'. *Wi: Journal of Mobile Culture*, 8(1). Retrieved from http://wi.mobilities.ca/mimi-sheller-mobility-justice.

Sheller, M. (2012). 'The islanding effect: Post-disaster mobility systems and humanitarian logistics in Haiti'. *Cultural Geographies*, 20(2), 185–204.

Soja, E. W. (2000). *Postmetropolis: Critical studies of cities and regions.* Oxford: Blackwell.

The Critical Engineering Working Group (Oliver, J., Savičić, G. & Vasiliev, D.). (2011). *The critical engineering manifesto.* Berlin, DE: The Critical Engineering Working Group.

Urry, J. (2007). *Mobilities.* Cambridge: Polity Press.

Urry, J. (2003). *Global complexity.* Oxford: Polity Press.

Winner, L. (1980). 'Do artifacts have politics?' *Daedalus*, 109(1), 121–136.

Yaneva, A. (2017). *Five ways to make architecture political: An introduction to the politics of design practice.* London: Bloomsbury.

Yaneva, A. (2009). 'Border crossings. Making the social hold: Towards an actor-network theory of design'. *Design & Culture*, 1(3), 273–288.

9 Emergent and integrated justice

Lessons from community initiatives to improve infrastructures for walking and cycling

Denver V. Nixon and Tim Schwanen

Introduction

In this chapter we propose that conceptions of transport and mobility justice as developed in recent academic literature need to be developed further if they are to capture the spatiotemporal complexity of what justice is and means in the context of everyday urban mobility. This proposition is informed by our ongoing research on community initiatives seeking to provide or enhance infrastructures for cycling and walking in the cities of London and São Paulo. Interviews with initiative leaders highlight the multi-scalar, emergent and extemporaneous nature of ideas about justice that animate their practices: both personal experiences of the unjust and broader visions of the good life, city, society and planet underpin those initiatives, and the specific injustices targeted by such initiatives at a given moment can, and often do, shift, often in response to changes in their urban social and political context. The emergent and extemporaneous character of what those initiatives take to be unjust and therefore demanding intervention is reflected in, and enabled by, most initiatives' experimental, agile and adaptive *modus operandi*, as well as their small size and interpersonal nature: encounters with the intersecting disadvantages experienced by participants and beneficiaries repeatedly trigger new perceptions, affects, thoughts and ambitions regarding justice among initiative leaders.

We take an inductive approach to our argumentation and, first, discuss excerpts from 72 semi-structured, one-to-three-hour interviews with leaders, staff and beneficiaries of community initiatives focused on walking and cycling infrastructures as well as with staff from government and not-for-profit organizations in London and São Paulo. We then explore to what extent conceptions of transport and mobility justice in the academic literature correspond to those held by our interview participants before introducing some conceptual resources for extending thinking on mobility justice in ways that resonate with our empirical findings.

Justice in practice

The charities, cooperatives and social enterprises we investigated provide 'soft' and 'hard' infrastructures (Tonkiss, 2015): cycle ride training, cycle maintenance and repair training, group walks and rides, street furniture, safety equipment, road paint and signage, aesthetic improvements to pathways and staircases, street closures and paper or virtual navigation. Many initiatives cater specifically to the needs of those with disabilities, members of cultural and religious minorities, refugees and asylum seekers, women and gender-variant people, low-income residents of deprived neighborhoods, children and the elderly. Most run their initiatives in economically and/or racially marginalized neighborhoods, typically in peripheral areas of London and São Paulo. For instance, one initiative takes mostly women from a low-income *bairro* in peripheral São Paulo on group walks through their neighborhoods that are otherwise considered too dangerous for walking.

On a more general level, however, the examined initiatives suggest a temporally and spatially open path to achieving justice, owing to (a) their multiscalar visions (the interpersonal experiences of injustice and broader global visions that often, in combination, led individuals to start the initiatives and continue to inspire and guide their work) and (b) their temporally dynamic practices (the extemporaneous and improvisational approach to active transport infrastructure that facilitates ongoing adaptations and decisions on what is just in any particular moment).

The individual, their community and global visions

A majority of initiatives were started after their leaders personally or vicariously experienced disadvantage in everyday life. Consider, for instance, Peter,[1] the founder of a London-based organization that offers bicycles, cycle maintenance training and rider skills training for refugees and asylum seekers. His experiences were described by Susan, a staff member of the initiative Peter founded, as follows:

> So, Peter was mentoring a refugee when he was at university, and [the] said refugee was housed … quite far out of the center of London, and as a refugee, you have a lot of appointments to attend, and he was finding it really difficult to do that on the amount of money he had to live on. So Peter found a bike and fixed it up and gave it to him, and then [he] realized that that could probably help quite a lot of other people in a similar situation, and so he started it, informally and voluntarily, running it out of his garden, and then kind of grew it from there.

While Peter was not a refugee, an interpersonal and empathetic experience with someone who was inspired the creation of the organization that, with the help of paid staff and volunteers, has given away several thousand bicycles and cycle training to people in similar precarious circumstances since 2013.

Harold, the founder of a bicycle repair workshop and group cycling activities in a disadvantaged and culturally diverse neighborhood of outer London, was influenced by both positive and negative personal experiences:

The background of it is me as an individual. A child born after World War II in Britain of Afro-Caribbean parentage. ... As a child, you identify with certain things, and one of the things I identified growing up was that sense of community, and being able to knock on somebody's door on your road. ... You know, Uncle John. We all knew him as "uncle". He was white, but he was my uncle, and nobody could ever even dream about telling me that he wasn't my uncle. ... So one day my daughters ask me to go on a bike ride. I said, "Okay, let's get the bikes out the garage and let's go". A couple of young people, they said, "Can we come?" So of course, leaning to the kind of background that I grew up in, I couldn't say no ... there was a couple of them that didn't have a bike, they came on skateboards. So that's how it all was born. That's all, you know, because I thought that was unacceptable in 2005, and you have no bike ... living in one of the most advanced countries in the world, or so I'm told.

This founder felt positively influenced by what he perceived as a strong sense of neighborhood community that transcended color, and by the support it provided during his youth. He later felt obliged to cultivate this sense of community in a different London space-time where he saw a vacuum of this type of support, particularly with respect to youth mobility in a supposedly progressive part of the world.

Leaders and staff also had more complex and sometimes larger-scale visions of enhanced mobility, stronger community connections, better health and cleaner environments, sometimes of global proportions. Alton, a co-leader and partner mechanic of Harold, is a good example:

I said if we want this organization to lay foundation blocks where our community can take their rightful place within the human race ... . So I have visions of big global things, yeah? ... How can you change hundreds of years of people's attitude and what they're thinking, how can you do that? And our reply would be, by action.

The co-leader connects what he sees as his community with the larger idea of a global humanity that is sometimes exclusive.

Although Felipe started group rides to benefit the people of the São Paulo peripheral zone where he was from, a permanent injury suffered by a friend while cycling galvanized his decision to do so:

My involvement with activism was on the week, the day that Guilherme lost his arm on [a major boulevard].
O meu envolvimento com o ativismo foi na semana, no dia em que o [Guilherme] perdeu o braço na [omitted for anonymity].

But he and a co-leader also recognized the broader impact of their activities:

[Pedro] But I think that we act locally and it has had an impact ...
[Felipe] It had a global impact ...
[Pedro] Mas acho que é isso a gente agir localmente e teve um impacto ...
[Felipe] Teve um impacto global ...

Many of the grassroots organization leaders discussed these multi-scalar influences, from individual epiphanies to desires for community empowerment to shared responsibility for issues of global proportions.

Emergent adaptations and experimentation

Most of the organizations that ran initiatives were in a constant state of change. This was partly due to unreliable funding sources, but also because leaders and staff were constantly re-visioning and adapting their practices to suit the needs of diverse and ever-changing beneficiaries, and because understandings of what required intervention shifted.

The generally small numbers of staff and local orientation of initiatives provide greater flexibility than in larger voluntary sector organizations or government. Most organizations were run by a leader, staff and/or volunteers numbering no more than seven people, although one or two people was more common. Numbers of beneficiaries varied greatly, from roughly 25 to thousands of people per year, although most organizations worked with a portion of 'regulars', some of whom went on to become staff. Most of these organizations assist particular local communities of people (e.g., refugees, women, youth), and leaders shape their organizations to address the more complex or intersectional disadvantages found among ever-changing beneficiaries.

For instance, the organization mentioned above that caters to immigrant and asylum seeker cycling mobility in London found that they weren't getting many women beneficiaries, so they adapted:

> Someone who used to work here went around to some other refugee service centers and spoke to the women who are refugees and asylum seekers there and asked them, you know, "can you ride a bike, and would you ride a bike, and why can't you ride a bike?" And you know, generally it was because they came from places which were quite conservative, and, um, cycling as a woman hadn't been, either it hadn't been a priority or it hadn't been okay. ... So we got specific funding to run sessions where we teach female refugees to ride, and kind of try to level that playing field a little bit I suppose?
>
> (Susan, London)

This organization's sensitive and interpersonal approach revealed the need to create separate sessions for women refugees, thereby furthering visions of a more just mobility landscape. Similar changes were evident from interviews with cycling-oriented initiatives in London that specifically targeted children and youth.

An organization in São Paulo that aims to render the city's many staircases more user-friendly tailored one initiative in a low-income area to the needs of local schoolchildren, their parents, neighbors, local soccer groups and graffiti artists:

> I think the second intervention we did was the most interesting one. It was in the peripheral area ... close to a slum, like very informal urban tissue and everything. ... So, but for us, it was a very interesting project because the staircase was in front of a school, public school. So we engaged with the students and the [teachers] and the principal also because we ... we researched the use of the staircase, and [it] was mainly for access to the school. Students and parents are the main users, so it was important to build this project with them because they are the main users. ... It was very interesting because the school was very excited with this project and also some community groups like graffiti groups and soccer [football] groups in the neighborhood also were engaged with the project.
>
> (Francisca, São Paulo)

This staircase initiative was a product of ongoing dialogue with and active participation of local people. It was not an 'out of the box' application of formal 'best practice' planning policy, but rather an emergent and unique complex of improvements. This action reaped notable community benefits, with the percentage of residents perceiving the staircase as unsafe dropping from 53% to 31% and the number who found it a pleasant space increasing from 13% to 63%. Besides the progressive adaptations made in this particular initiative, the organization was later confronted by the need to rework, yet again, their approach to justice when asked by a developer to improve a staircase next to a proposed condominium building in a lower-middle-class, but gentrifying, area of the city.

The diversity of *modus operandi* in these examples resembles bicycle repair workshops in Batterbury and Vandermeersch's (2016) study and reflects an emergent or extemporaneous form of justice. In this way, what is seen as just and in need of action changes as organizations' practices and the makeup of the beneficiaries change across time and space. In fact, our findings suggest that theoretical constructs that help researchers think about transport and mobility justice should meet two conditions: enable researchers to consider simultaneously multiple spatial scales of action and impact, from the individual to all life on earth; and draw attention to the value of extemporaneity in process or procedure.

Justice in theory

There is growing interest in justice in transport and mobilities research, and in this section we examine if and how the burgeoning literature in these fields satisfies the two conditions inductively derived from our empirical research.

Transport justice

Perhaps the oldest and most voluminous strand of justice research in relation to transport and mobilities is concerned with uneven distributions of motility – the capacity to move – and particularly accessibility – the ease with which destinations where basic needs can be fulfilled and wants pursued can be reached (Wachs & Kumagai, 1973). The efforts of researchers like Mei-Po Kwan (e.g., Kim & Kwan, 2003; Kwan, 1998; Kwan, Murray, O'Kelly & Tiefelsdorf, 2003) have led to more sophisticated modeling of the distribution of accessibility that takes into account space and time with a finer granularity. However, the emphasis on empirical, often quantitative, assessments of uneven distribution overshadows consideration of how and by whom these distributions are *processually* created and maintained (Cook & Butz, 2016; Schweitzer & Valenzuela, 2004); such assessments cannot capture David Harvey's (1973, p. 8) understanding of social justice as "the just distribution justly arrived at".

Within transport justice discourse, Martens (2016) proposes that transport – much like healthcare or education – requires its own distributive principles because it consists of a specific set of essential goods with (spatially contingent) shared meanings. Those principles should be based upon thresholds, as distribution cannot be left to market-based demand. In a related vein, Pereira, Schwanen and Banister (2017) suggest, after Rawls (1971), that accessibility and government investments supporting it should be considered primary goods and that establishing minimum thresholds of transport accessibility would be the best way to realize a more just society. Yet, both approaches, which are rooted in liberal political philosophy, are premised on static conceptions of what is distributionally and procedurally just. While recognizing spatial heterogeneity and societal contingency of needs and capabilities, both follow Rawls in assuming that traditional political process will generate ongoing consensus, with minimal contestation, on what precisely just distributions and procedures are (Sokoloff, 2005). The determination of accessibility thresholds is also problematic spatially because it risks unjust distributions of transport impacts across different geographical contexts. For instance, Pereira, Schwanen and Banister (2017) suggest that following a Rawlsian policy approach to justice in a low-density or rural area might see the facilitation of car use for low-income people; this may distribute local accessibility more equitably, but what would this expansion of the automobile fleet in the rural United States mean for citizens of the Marshall Islands given the contribution to anthropogenic climate change impacts?

Mobility justice

Mobility justice is a relatively new synthesis of mobilities and justice thinking. It explicitly applies ideas from critical social sciences to questions around mobility and justice. Sheller (2015, 2016) juxtaposes high-level assessments of 'transport justice' (e.g., national distributional statistics) with the potential of 'mobility justice' to offer a nuanced analysis that is sensitive to justice-related processes of culture and history. She demonstrates, for instance, how both a Foucauldian genealogical analysis of power and the multilevel perspective on transitions can improve understandings of uneven mobility. Because this means that cultural history and niches, regimes and landscapes are simultaneously scrutinized, Sheller's work represents an expansion of the spatial scales and temporalities considered, in contrast to transport justice's more static and universal conceptions of justice.

In their work on the Attabad landslide in Pakistan, Cook and Butz (2016) call for a departure from strictly distributive analyses of mobility justice and a turn towards histories of governmental regimes of power and domination that create the infrastructural and sociopolitical contexts that exacerbate injustice during and after disasters. This too represents a turn toward longer timescales and the connection between communities and the larger spatialities of governance.

Despite these helpful expansions of the temporal and spatial dimensions of mobility justice, both Sheller (2015) and Cook and Butz (2016) understand the production of mobility infrastructure and (in)justice as centered primarily within the governance bodies of regimes. This overlooks the justice-related processual agency, beyond simply votes or lobbying, that may be found at the level of individuals, communities or niches.

The collective right to mobility

Other scholars have detailed what might be characterized as a collective approach to mobility justice, where most often urban communities, collectives and coalitions seek a right to mobility. Although the 'right to mobility' was first described by scholars interrogating international migration (e.g., Cresswell, 2006; Sheller, 2008), Verlinghieri and Venturini (2018, p. 127) have synthesized the idea with Lefebvre's right to the city using the concept of access, defining it as "the right to move in the urban space, to access places and opportunities, but also the right to stay still". Like Lefebvre's right to the city and Harvey's (2008) interpretation of it, the right to mobility also includes the collective right to participate in decisions that lead to its realization (Verlinghieri & Venturini, 2018). In his research on the LA Bus Riders Union, Soja (2010) emphasizes the importance of building strong coalitions of communities in order to enact this participation.

However, despite Harvey's and Lefebvre's foregrounding of the influence of global capital on spatial rights, and Soja's elaboration of three spatial levels of spatial justice, the right to the city or mobility remains grounded in an urban context. Besides questions around what precisely defines the urban, this limits the spatial scale of justice inquiry and perhaps falls into the 'local trap' – the assumption that the local scale is more politically potent than other scales of citizen engagement. There is no evidence that this is the case; rather, all scales are associated with both advantages and disadvantages with respect to political engagement (Purcell, 2006, p. 1921).

Similarly, what is meant by a collective or community is often left undefined. Aiken (2015) argues that the term 'community' is polysemous and that its deployment often assumes an easily circumscribable group that possesses a certain level of homogeneity in the aims and values of its members. There are many historical examples of very effective collectives of the politically engaged who heaped deep injustices on those outside the collectives. In these cases, the gut feeling of a dissenting *individual* may play an important role in breaking harmful regimes. Also, injustices are often products of diverse and intersecting disadvantages operating across multiple scales and changing over time. The bluntness of community aggregations may miss this complexity. Finally, the analytical replacement of the individual with the collective may risk the dismissal of the situated, subjective, embodied *person* who, through the conscious or unconscious sediments of past interactive experiences, retunes her practices (Parr, 2010).

Moreover, some notions of 'rights' themselves may be problematic because of their abstract quality, which masks the spatially situated and demographically uneven efficacy of rights claims and assumes unanimous agreement on the shape of the collective right to a homogenous conception of mobility (Cresswell, 2006; Pratt, 2004). This also ignores complications like the "paradox of freedom", whereby one person's or group's freedoms, or rights, may intrude upon another's (Popper, 1966). The abstract nature of rights was one of the reasons some of our participants in São Paulo saw the right to the city as a concept demographically exclusive to highly educated people – they claimed the average person could not relate to the idea (see also Merrifield, 2011) – and limiting the breadth of discourse on pedestrian and cycle mobility benefits. However, the right to the city is not conceptualized as a legal right (Attoh, 2011; Harvey, 2008) and, thus, may not be susceptible to the same level of abstraction as found in law.

In summary, the commons approach to mobility justice is helpful in the way it brings in a politics of process, but it does not go far enough. We cannot say that any of these theories are, or are not, *just*, but we must recognize the complexity of mobility justice practices found everywhere 'on the ground'. Placed beside the empirical examples of grassroots justice practices presented earlier, we argue that these conceptualizations would benefit from meeting the conditions described earlier; namely, the simultaneous consideration of multiple spatial scales of (un)just action and impact, and a less static notion of justice.

Toward spatially and temporally complex conceptualizations

We argue that there is a need to bring in more relational, multi-scalar and emergent perspectives on mobility justice. Here we propose that conceptual resources for advancing such perspectives can be found in both Western and Eastern philosophies.

The very boundless, dynamic and unpredictable nature of social and environmental impacts and distributions of mobility means that justice must entail globally scaled considerations. An ethically strong version of mobility justice must include "justice beyond the human" (Schlosberg, 2013, p. 40), including animals and plants. Climate change is a powerful example of the importance of an expanded scope of justice. Consider, for instance, the effects of US automobility on climate change, and the partial effect of climate change on landslide frequencies in the mountainous areas of South Asia (Crozier, 2010). At this point in time, transport and mobility justice is focused perhaps too strongly on the distribution of motility, accessibility and mobility as goods and their procedural reproduction, thereby marginalizing the potentially (un) just global impacts they may have.

However, the intersectional, interpersonal and sometimes locally specific nature of mobility injustice demands concomitant individual and community perspectives on, and approaches to, those dynamics and solutions to them. Abstract notions of a better world are, after all, most provocative when co-constituted with personal or collective sense experience. Merrifield (2011, p. 478) describes the potential experiential change agent as "waiting for something closer to home, something trivial – something he [sic] can touch and smell and feel – *and* for something larger than life, something that's also world-historical". Batterbury and Vandermeersch (2016, p. 198) similarly claim that mobility justice will only be effective when it "acknowledge[s] the lived experiences of transport users", and Nixon (2014) argues that embodied experience with multiple transport modes cultivates empathy for users across those modes. Coming together through embodied encounters in specific places may be more effective in igniting this collective power than an abstract rallying call to a right to the city (Merrifield, 2011).

How should global, collective and individual scales of justice be merged? Daoism is a Chinese philosophy *and* religion (Schipper, 1993) that is approximately 2,500 years old.[2] The fanciful narrative in the first chapter of the canonical Daoist text *Zhuangzi* plays with the reader's sense of space and scale (Shen, 2009), reflecting Purcell's (2006) claim, over two millennia later, that scale is socially produced rather than an ontological given. In this way, Daoism problematizes the division of experience into distinct categories of 'reality', including the spatial categories of scale.

But perhaps most relevant here is the spatial application of philosopher Roger Ames' interpretation of '*dao*' and '*de*', the same terms in the title of the *Dao De Jing* (Tao Te Ching). Ames proffers that *dao* represents (ever inadequately) a cosmological and ontological undivided 'whole', whereas *de* represents a singular particle that might be akin to an individual were it not for its inseparability from the *dao*. He also relates the *dao* and *de* to Plato's 'abstract universal' and 'concrete particular', respectively. Ames (1986, p. 342) suggests that the "*[d]ao* be understood as an emerging pattern of relatedness perceived from the perspective of an irreducibly participatory *[d]e*". In other words, we cannot pretend that individuals exist as atomistic entities, but nor can we ignore the unique diversity of the 'myriad things', including perspectives, that are constantly in transformative relationships within the ever-emerging whole.

Applying this idea to the mobility justice problematique, considering individuals as distinct, atomistic entities is folly, as is forgetting the particulars that constitute universals. Balancing and integrating particularity (e.g., individual people) and generality (e.g., notions of 'the world') with things at intermediate scales (e.g., groups or nation-states) constitutes a more holistic and ultimately just approach to mobility justice. Thinking in this manner starts from the idea that, for example, the importance of a particular individual to a justice movement, local (un)just action by that movement and the role of future generations of people, animals or plants on the other side of the world are deeply entwined and co-constitutive of each other. This conception of space satisfies our first condition of transport and mobility justice, mentioned above, as it simultaneously attends to multiple spatial scales of action and impact.

The second condition calls for a more complex conception of time. The temporality of transport and mobility justice needs to be rethought in three related ways: first, more attention must be paid to (un)just relationships between generations; second, there is a need to consider the broader potential futures that may emerge from present action; and third, the dynamism and emergence of (un)just relationships must be recognized. Whereas existing formulations of mobility justice include historical considerations, they do not mention intergenerational injustice and, with the exception of the commons approach, are largely based on relatively static, or at least slow-moving, justice procedures such as policy and planning. In addition to considering the past, mobility justice must be mindful of potential futures. As seen in the practices of the community initiatives focused on walking and cycling infrastructures, it also requires a more emergent conception of the present to expand understandings of procedural justice. Derrida and Daoism are helpful here.

Procedural (in)justice is often thought to be ossified in transport policy and planning documents, or entrenched in the practices of hegemonic regimes, such as national governments that 'only know how' to build airports for the kinetic elite. However, the creation of just or unjust mobility infrastructure may also be seen as an ongoing negotiation that must be "sensitive to the specificity of each moment" (Sokoloff, 2005, p. 343) and is constantly in a state of existential 'becoming' (Derrida, 1994, p. 175). Within Daoism, there is a long tradition of people and groups cultivating an adaptive, extemporaneous way of being in the world (Nixon, 2006) that "eschew[s] antecedent principles or norms" (Hall, 1987, p. 170). That is, Daoists prefer to adaptively respond to the 'flux' of life 'in the moment', without prejudice or a preconceived set of rules of engagement. That is not to say that Daoists are immune to influence from past experiences, but they attempt to let go of assumptions through mental and physical exercises and engaging with the world. Adopting this orientation when confronting transport and mobility injustices allows for a more flexible and adaptive form of procedural justice practice to develop.

We argue that by adopting emergent sensitivity, rather than predetermined static rules, in the practices of government and community-led infrastructure provision initiatives, more just mobility outcomes are possible. This extemporaneous approach may involve increased and ongoing community inclusion in decision-making processes, greater interpersonal relationships among community members, a more experimental approach to infrastructure or a range of other possibilities.

Conclusion

This way of thinking about justice – in which particularity and generality are integrated, from the individual to world scale, and static principles are replaced by an emergent practice that is contingent on who participates and what unfolds – can be generalized beyond the empirical cases discussed at the beginning of this chapter. Take, for example, an electric car club that provides a car to a paraplegic member who is forced by economic circumstance to live in a suburb. It may be a more just mobility option than others, at a particular point in time, for the club member, but also for the homeless person who otherwise is exposed to fumes in the underground parking lot where the club member regularly shops. However, introducing a local not-for-profit that provides specialized cycles to those with disabilities along with a regime shift that renders roads safer for cyclists (e.g., speed limit reductions, more separate facilities and stronger law) may open up better options for both the car club member's health and climate change impacts felt on the other side of the world. Also, different people may have different views on how best to confront issues of mobility justice such as this one, and thus dialogue is necessary. Justice practiced in this emergent, multi-scalar way has the potential to detect new injustices that arise over time or that exist at a scale overlooked in prior justice efforts.

To develop the ideas put forth here, different, somewhat unorthodox ways of speaking about mobility justice may be necessary. Research into mobility justice may also do well to explore practices of justice outside the bounds of government procedure, such as efforts from within civil society, as shrinking budgets and meso-spatial levels of responsibility often limit governance to the pursuit of general and enduring principles of justice dispensed in a 'bulk' fashion.

This expanded conceptualization of mobility justice may help planners and other decision makers: (a) recognize the role that civil society organizations – with their ongoing adaptations and multi-scalar considerations of justice – can play in bringing justice to the mobility landscape, particularly when their impacts are aggregated; (b) create greater awareness about the connections between individuals in their communities and injustices that manifest at greater scales; and (c) rework planning and operational procedures such that transport or mobility justice is seen as an unfinished work in need of ongoing discussion with both the communities for which the representatives are formally responsible and others, such as the people in other places who may be affected by these plans and procedures.

The value of community initiatives focused on walking and cycling infrastructure lie less in what they accomplish infrastructurally in an absolute sense, and more in what they represent as a 'living', dynamic, processual and multi-scalar example of collaboration, connection, sensitivity and experimentation in the interest of achieving a just society. They also demonstrate the value of more spatially plural conceptions of justice.

Notes

1 All names are pseudonyms.
2 We use the Pinyin transliteration of Chinese here, rather than the older Wade-Giles. Thus, what was transliterated as "tao" and "te" becomes "dao" and "de".

References

Aiken, G. (2015). '(Local-) community for global challenges: Carbon conversations, transition towns and governmental elisions'. *Local Environment*, 20(7), 764–781.

Ames, R. T. (1986). 'Taoism and the nature of nature'. *Environmental Ethics*, 8(4), 317–350.

Attoh, K. A. (2011). 'What kind of right is the right to the city?' *Progress in Human Geography*, 35(5), 669–685.

Batterbury, S. & Vandermeersch, I. (2016). 'Bicycle justice: Community bicycle workshops and "invisible cyclists" in Brussels'. In A. Golub, M. L. Hoffman, A. E. Lugo & G. F. Sandoval (Eds.), *Bicycle justice and urban transformation: Biking for all?* (pp. 197–209). Abingdon, UK: Routledge.

Cook, N. & Butz, D. (2016). 'Mobility justice in the context of disaster'. *Mobilities*, 11 (3), 400–419.

Cresswell, T. (2006). 'The right to mobility: The production of mobility in the courtroom'. *Antipode*, 38(4), 735–754.

Crozier, M. (2010). 'Deciphering the effect of climate change on landslide activity: A review'. *Geomorphology*, 124(3–4), 260–267.

Derrida, J. (1994). *Specters of Marx: The sate of the debt, the work of mourning and the new international*. (P. Kamuf, Trans.). New York: Routledge.

Hall, D. L. (1987). 'On seeking a change of environment: A quasi-Taoist proposal'. *Philosophy East & West*, 37(2), 160–171.

Harvey, D. (2008). 'The right to the city'. *New Left Review*, 53, 23–40.

Harvey, D. (1973). *Social justice and the city*. London: Edward Arnold.

Kim, H. M. & Kwan, M. P. (2003). 'Space–time accessibility measures: A geocomputational algorithm with a focus on the feasible opportunity set and possible activity duration'. *Journal of Geographical Systems*, 5(1), 71–91.

Kwan, M. P. (1998). 'Space–time and integral measures of individual accessibility: A comparative analysis using a point-based framework'. *Geographical Analysis*, 30(3), 191–216.

Kwan, M. P., Murray, A. T., O'Kelly, M. E. & Tiefelsdorf, M. (2003). 'Recent advances in accessibility research: Representation, methodology and applications'. *Journal of Geographical Systems*, 5(1), 129–138.

Martens, K. (2016). *Transport justice: Designing fair transportation systems*. London: Routledge.

Merrifield, A. (2011). 'The right to the city and beyond: Notes on a Lefebvrian reconceptualization'. *City*, 15(3–4), 473–481.

Nixon, D. V. (2014). 'Speeding capsules of alienation? Social (dis)connections amongst drivers, cyclists and pedestrians in Vancouver, BC'. *Geoforum*, 54, 91–102.

Nixon, D. V. (2006). 'The environmental resonance of Daoist moving meditations'. *Worldviews: Global Religions, Culture & Ecology*, 10(3), 380–403.

Parr, J. (2010). *Sensing changes: Technologies, environments and the everyday, 1953–2003*. Vancouver, BC: University of British Columbia Press.

Pereira, R. H., Schwanen, T. & Banister, D. (2017). 'Distributive justice and equity in transportation'. *Transport Reviews*, 37(2), 170–191.

Popper, K. R. (1966). *The open society and its enemies* (Vol. 2). Princeton, NJ: Princeton University Press.

Pratt, G. (2004). *Working feminism*. Philadelphia, PA: Temple University Press.

Purcell, M. (2006). 'Urban democracy and the local trap'. *Urban Studies*, 43(11), 1921–1941.

Rawls, J. (1971). *A theory of justice*. Cambridge, MA: Harvard University Press.

Schipper, K. (1993). *The Taoist body*. Berkeley, CA: University of California Press.

Schlosberg, D. (2013). 'Theorizing environmental justice: The expanding sphere of a discourse'. *Environmental Politics*, 22(1), 37–55.

Schweitzer, L. & Valenzuela, A., Jr. (2004). 'Environmental injustice and transportation: The claims and the evidence'. *Journal of Planning Literature*, 18(4), 383–398.

Sheller, M. (2016). 'Uneven mobility futures: A Foucauldian approach'. *Mobilities*, 11 (1), 15–31.

Sheller, M. (2015). 'Racialized mobility transitions in Philadelphia: Connecting urban sustainability and transport justice'. *City & Society*, 27(1), 70–91.

Sheller, M. (2008). 'Mobility, freedom and public space'. In S. Bergmann & T. Sager (Eds.), *The ethics of mobilities: Rethinking place, exclusion, freedom and environment* (pp. 25–38). Oxford: Ashgate.

Shen, V. (2009). 'Zhuang Zi and the Zhuang Zi'. In B. Mou (Ed.), *History of Chinese Philosophy* (pp. 237–265). London: Routledge.

Soja, E. (2010). *Seeking spatial justice*. Minneapolis, MN: University of Minnesota Press.

Sokoloff, W. (2005). 'Between justice and legality: Derrida and decision'. *Political Research Quarterly*, 58(2), 341–352.

Tonkiss, F. (2015). 'Afterword: Economies of infrastructure'. *City*, 19(2–3), 384–391.

Verlinghieri, E. & Venturini, F. (2018). 'Exploring the right to mobility through the 2013 mobilizations in Rio de Janeiro'. *Journal of Transport Geography*, 67, 126–136.

Wachs, M. & Kumagai, T. G. (1973). 'Physical accessibility as a social indicator'. *Socio-Economic Planning Sciences*, 7(5), 437–456.

10 Fighting for ferry justice

Sharon R. Roseman

Introduction

This chapter examines the consolidation of a language of mobility justice during a specific historical moment in Bell Island, Canada, between the 1950s and 1980s. Bell Island's iron ore mining industry slowed down and then closed in 1966, forcing many workers to undertake a daily off-island commute on a public passenger and vehicle ferry service to access new jobs on the large island of Newfoundland.[1] The ferries travel five kilometers in Conception Bay, connecting Bell Island to the Newfoundland town of Portugal Cove-St. Philips, under 30 minutes' drive to the provincial capital St. John's.[2]

By the early 1970s, the main channel for Bell Islanders' calls for mobility justice was the grassroots Bell Island Commuters' Committee. Later renamed the Bell Island Ferry Users Committee, it continues to play a mobility advocacy role today. Suzan Ilcan (2013, p. 4) notes that "[t]he mobility-knowledge nexus has crucial implications for social justice". For more than four decades, Bell Island's ferry advocacy committees have drawn on an in-depth local knowledge of users' experiences and needs to formulate strategic commentaries. As David Bissell (2015, pp. 160, 161) outlines, commentaries about specific forms of mobility – whether articulated by grassroots advocacy groups, politicians, civil servants or individual commuters – comprise "a constitutive aspect of the event of commuting" and can be "key micropolitical agent[s] of transformation".

Bell Islanders' ferry committees make the case that access to mobility infrastructures is a key component of mobility justice. They highlight how mobility injustices occur as a result of insufficient capacity on boats, inadequate scheduling, mechanical problems, ticket costs, physical barriers to accessibility and other hindrances. The committees demonstrate how gaps in the availability of affordable, accessible and dependable modes of transportation lead to "transport-related social exclusion", including difficulties securing paid employment, healthcare and education (McDonagh, 2006, p. 355; see also Hine, 2011). Despite their relative proximity to a capital, Bell Islanders' experiences can be compared broadly to spatialized "transport inequity" found in other rural areas, islands and remote regions (Sheller, 2011, p. 293;

see also Cook & Butz, 2010, 2016, p. 401; Velaga, Beecroft, Nelson, Corsar & Edwards, 2012).

I employ Henri Lefebvre's (1968, 1991, 1995) ideas about economically and politically marginalized citizens' 'right to the city' and extend these to those living in peri-urban, rural and remote regions. As Chris Butler (2012) argues, Lefebvre emphasizes that all city dwellers should be able to participate in urban space writ large, as opposed to being confined to specific neighborhoods as a result of class hierarchies and other unequal social relationships. This focus on a right to the city coincides with the attention Lefebvre pays to the urban citizen as a figure of potentially transformational politics (see also Harvey, 2008; Purcell, 2002; Shields, 2013; Staeheli, Dowler & Wastl-Walter, 2002). In considering the prospective agency of this political figure, Lefebvre (1991, p. 386, as cited in Purcell, 2014, p. 145) argues that change needs to include "political sites beyond the workplace" and the electoral system. Bell Islanders' fight for a better ferry service fits Lefebvre's vision. They have not "surrender[ed]" "collective decisions" about their mobilities "to a managerial class" – in this case the officials responsible for the provincial ferry services (Purcell, 2014, p. 147, drawing on Lefebvre, 2003[1970]). For over four decades, they have struggled against explicit and implicit pressures either to abandon their island home or to continue contending with insufficient ferry service. Their calls for ferry justice assert a right to the city and, by extension, to the aquamobile transportation required to move them to and from St. John's and other places. In maintaining residences on Bell Island while accessing jobs and basic services in St. John's, Bell Islanders are defending their right to the city *and* their right to live on an island. They are arguing for nonurban localities, including small islands, as counter-spaces with respect to normalized urban-centric spatial arrangements. Similarly, they defend ferries as counter-mobilities vis-à-vis the cultural dominance of land- and air-based travel (Lefebvre, 1991, pp. 349, 367, 381–383; also see Schwartz, 2014). 'Counter-space' (Lefebvre, 1991, p. 349) is a play on words referring to a "space of counter-culture" in which people resist the social status quo by asserting reimagined alternatives. Fights for mobility justice in the form of improved access to public transit can be understood as explicit counters to hegemonic mobilities systems, which are often entrenched in spatial arrangements that disadvantage particular groups.

As Batterbury (2003, p. 152) outlines, "[u]rban social movements can, and frequently do, arise around urban transportation issues". Take, for example, the TTC Riders,[3] founded in 2010 in Toronto, Canada's most populous metropolitan region. This organization argues that "public transit" is a "social good" akin to public education and healthcare (Huang, 2014, p. 17; also see Hulchanski, 2010). In the Newfoundland and Labrador urban context, grassroots activism calls for improved public transportation through ameliorated bus service, the addition of cycling lanes and access to all-year pedestrianism achieved through winter sidewalk snow-clearing (Yeoman & Roseman, 2016). Local residents' calls for ferry justice in Bell Island and other aquamobile-reliant locations in Newfoundland and Labrador highlight the importance of considering the similar and often longstanding role of mobility justice activism outside cities.

The following section provides an overview of the historical context for the mid-20th-century emergence of a specific language of mobility justice in Bell Island. I next turn to two milestone documents prepared by the ferry advocacy committees in the 1970s and 1980s that developed and reinforced a message regarding Bell Islanders' right to St. John's and their resistance to mobilities regimes that would depopulate rural and island communities, and I then reflect on the continued salience and relevance of mobility justice assertions in this context.

Fighting for a right to the city

Similar to other island and coastal communities reliant on aquamobility (e.g., Vannini, 2011, 2012), Bell Islanders' political sensibilities have been shaped by a ferry service that comprises their only public transportation link with the rest of the province. The 1966 shuttering of the island's one remaining submarine iron ore mine ended a 70-year history of mining as a source of direct and indirect local employment (Weir, 2006). In considering this event's mobility implications, it is useful to take a "regimes-of-mobility" approach, which foregrounds the political economic conditions that produce differential access to forms and conditions of mobility in specific historical settings (Glick-Schiller & Salazar, 2013, p. 195; see also Cresswell, 2010; Kesselring, 2014). A mid-20th-century reduction in mining employment in Bell Island preceded the final closure and exacerbated then-existing ferry service deficiencies related to capacity, consistency and scheduling. A private firm, the Newfoundland Transportation Company, held an exclusive operating franchise for this ferry route since 1955, supported through public subsidies.[4] In 1960, the federal government commissioned the new ferry MV *John Guy* specifically for Bell Island. However, it did not pay sufficient attention to the wharf modifications required to accommodate the new ferry. In September of that year a committee of five men met with the provincial premier, Joseph R. Smallwood, to discuss a need for better docking facilities for the new boat (*The Daily News*, 1960).

Transportation and the right to the city were at the forefront of civil society reactions to the employment crisis that followed several years later. In June 1966, the same month when No. 6 mine was shuttered, a 'citizens' committee' of 12 Bell Islanders met with Steve Neary, their Member of the provincial House of Assembly. They presented him with a petition calling for free ferry service as an affordable option for daily commuting to the St. John's region. Signed by 80% of eligible voters from the island, it was subsequently presented by Neary to Joseph R. Smallwood (*The Daily News*, 1966a, 1966b, 1966c). This petition followed earlier requests for schedule improvements, all oriented to ensuring Bell Islanders could move efficiently and affordably back and forth to jobs in the St. John's area (Gendreau, 1966).

The mobilities regime of the mine closure period coincided broadly with pressures in Canada for urbanization and the relocation of vulnerable (often rural) labor pools, pressures at times exerted by federal and provincial government policies and practices as well as produced by broader labor market shifts. Bell Island's population exceeded 12,000 by the mid-20th century, but declined dramatically after mining ended (Weir, 2006). As a result of political pressure on both levels of government to respond to the crisis, a federal-provincial committee was established to find solutions, and a door-to-door survey was conducted. 'Resettlement aid' funding was offered to Bell Islanders in the late 1960s through the federal Manpower Mobility Program in conjunction with other agencies (Atlantic Development Board, 1967). Even though some households in desperate economic straits welcomed this funding, the process is still criticized for having compelled families to make quick decisions about giving up their homes in exchange for a modest cash sum. The Manpower Mobility Program aligned with similar large-scale government resettlement programs in Newfoundland and Labrador in the same time period, which together resulted in the abandonment of many coastal settlements in Newfoundland (Martin, 2017[2006]). Resettlement mobilities reduced the cost of providing government services, such as schools, wharves and roads, and provided concentrated pools of labor for seafood plants and other employment points located in other rural and urban settlements.[5]

Despite government attempts in the 1960s and 1970s to coerce residents into leaving Bell Island, civil society pushback and support from politicians representing the area led both federal and provincial levels of government to acknowledge that the ferry service was still needed "for effective daily commuting to St. John's" (Atlantic Development Board, 1967).[6] The federal government had provided funding for some intra-provincial ferry routes, including Bell Island's, from 1949, when Newfoundland joined Canada. In 1979, the province agreed to gradually take over full fiscal responsibility for intra-provincial ferries (CNM Inc. 1981; Collier, 2010). Therefore, Bell Islanders communicated with both federal and provincial politicians and entities about ferry service in the period when mining ended.

In the lead-up to the formation of the local ferry committee, therefore, this transportation service was a central theme for Bell Islanders and others advocating for the fair treatment of displaced mine workers and their families and neighbors. Reduced fares for daily commuters were instituted in 1966. On November 30, 1966, the issue was discussed in the Canadian parliament, with Prime Minister Pearson responding to a question from the St. John's East Member of Parliament, Joseph O'Keefe, whose riding included Bell Island. Pearson noted that the interdepartmental federal government committee led by the Atlantic Development Board, which was formed to look into solutions, had "arranged, through the Canadian Maritime Commission, for lower ferry commuter rates between Bell Island and Newfoundland to assist the mobility of the Bell Island labor force" (House of Commons Debates, 1966; see also *The Evening Telegram*, 1966a, p. 3). Four days later, new fare rates for work commuters were announced (*The Evening Telegram*, 1966b, p. 4). In addition to the reduced fees, commuters gained priority ferry access for vehicles on the Bell Island side (Atlantic Development Board, 1967). During a December 8, 1966, debate in the provincial legislature, Steve Neary maintained that islanders should not have to pay ferry rates at all given the construction of non-tolled bridges, causeways and roads in other parts of the province that replaced a reliance on ferries (Guy, 1966, p. 3). Ray Guy, then a staff writer for the *Evening Telegram*, reported on this debate. He quoted Neary's comment that eliminating ferry fees "would be 'social justice' … [s]o that Bell Islanders might share the same advantage as other Newfoundlanders in being able to drive freely about the province" (Guy, 1966, p. 3).[7]

The Bell Island Commuters' Committee was established in the early 1970s following consistent calls by citizens and local officials for an improved service with increased subsidization and oversight of the private company operating the boats (Bell Island Commuters' Committee and Representative Community Groups, 1973). In the 1980s, its name was changed to the Bell Island Ferry Users Committee to reflect the range of individuals that relied on the ferry beyond those commuting for employment and education. In the same time period, the Newfoundland Transportation Company's franchise ended, and the provincial government took over the operation of the Bell Island service. The largely volunteer committee continues to be in place today, advocating for ferry justice in the face of often challenging circumstances.[8] Bell Islanders, including the dozens who have sat on ferry advocacy committees, have asserted their entitlement to influence public transportation services. In so doing, they have argued for their right to move back and forth to the city to access employment while continuing to live on an island that sits in close geographical proximity to the capital. Committee members, therefore, have long rejected the assumption – prevalent in both North America and Europe – that workers based in small towns and rural areas should move to urban centers and other regions without the social supports and other key resources available in their home communities (Roseman, 2013). Bell Islanders' assertions that they should have some say over their public ferry service are closely related to their overall efforts to sustain and revitalize their home island. These efforts are fundamentally tied to their resistance to various forms of social exclusion, including difficulties with reaching employment, education and healthcare as a result of an insufficient public transportation system. As is the case with rural communities elsewhere in the world, they are fully engaged in the making of counter-spaces and counter-mobilities (Lefebvre, 1991; Roseman, 2008).

"Mr. Premier, consider for a moment the plight of a workman": calls for mobility dignity

When the Bell Island Commuters' Committee was formed, a newspaper report summarized its purpose as being "to register grievances over the services" (*The Daily News*, 1972a, p. 3). In that regard, the committee's main emphasis was to ensure the possibility of daily commuting to jobs in the St. John's area. To achieve this goal, the committee appealed for federal and provincial government intervention into the operation of the private company with the tender for the service, to eliminate, reduce or control rates and to expand and intensify the ride schedules; calls were also made for an open public tendering of the service (Bell Island Commuters' Committee and Representative Community Groups, 1973, p. 5). Requests were made for better boats and wharves; more crew; better waiting rooms, parking and line-up areas; more affordable bus or taxi services in Portugal Cove; and the subsidization of contingency air transportation and lodging when the ferries could not operate for extended periods.

I discuss two milestone committee commentaries below. These documents develop the language of mobility justice as a response to an overarching mobilities regime that provided challenges to the continuing viability of specific rural and island communities. One of the central themes is an appeal for dignity for Bell Islanders who commute by ferry. The first commentary, from 1973, is a detailed brief "regarding the Bell Island Ferry System", submitted to the provincial premier, Frank D. Moores, by the Bell Island Commuters' Committee and Representative Community Groups. It followed a series of other formal commentaries presented by the committee in its early days (*The Daily News*, 1972a; 1972b). The second document is a 1983 brief, prepared by the committee when the entire provincial cabinet came to Bell Island, at the time when the province was taking over responsibility for intra-provincial ferries.[9]

The 1973 document develops the argument that Bell Islanders should have the right to remain in their island homes, to secure employment and to access daily aquamobility. The demand for conditions that would allow for mobility dignity for commuting workers is central. This commentary is framed to address the provincial leader personally, imitating personal correspondence sent to him and other politicians. Here is a striking passage:

> Mr. Premier, consider for a moment the plight of a workman who leaves his home on Bell Island in the morning at 6:45, begins work at his place of employment in St. John's or surrounding area at 8:00 a.m., works hard and industriously through the day, finishes at 5 or 6:00 p.m.; the same man returns to Portugal Cove at suppertime hoping to find accommodation on the ferry to Bell Island; he finds that the ferry has left for the Island at 7:00 p.m. and won't be making another crossing from Portugal Cove 'till 11:10 p.m.; or he may have arrived in time only to find that the ferry is unable to accommodate him and a number of his working cohorts with their vehicles, due to the volume of traffic. Nevertheless, he has to wait an extra 4 hours.[10]
>
> (Bell Island Commuters' Committee and Representative Community
> Groups, 1973, p. 3)

The brief goes on to emphasize that this imagined worker is not alone, because "approximately 150 heads of households commute daily, and another

150 do so weekly and/or bi-weekly for the same purpose", and "support approximately 20% of our present population (6500)" (p. 4). But "[s]urely they deserve better treatment!" (p. 4). Some "lost days wages or were 'late for work' due to the inability of the ferry JOHN GUY to accommodate them" (pp. 4–5). Furthermore, there were "instances when many of our people have had to stay for the night in the waiting rooms due to lack of accommodation on the ferry" (p. 5). Others were not able to work because their "job demands 'shift work'" (p. 4) and the last ferry left Portugal Cove at 7 p.m.

To mirror the style of personal address used in the brief, *The Daily News* (1973a, p. 2) reported on August 2 that "[a] delegation from the Bell Island Commuters' Committee meets with Premier Frank Moores at 10:30 a.m. tomorrow to present a brief ... the delegation wants to impress upon the provincial government the seriousness of the ineffectiveness of the service". In another news item published on August 31, readers learned that "[a] new approach to the Bell Island ferry service was announced ... [that] will provide for extra runs each day ... operating on a trial basis" (*The Daily News*, 1973b, p. 18).

The 1983 brief echoes and builds on the 1973 commentary about the need for justice and dignified commuting conditions, while also staking a claim to effective agency as a committee:

> Since 1972 there have been several significant improvements to our community, especially the Ferry Service ... the priority given to daily working commuters, stabilized ferry rates and extended scheduling (etc.) have certainly improved the social and economic prospects for our people. Our committee, volunteer and publicly elected from the Community, are most pleased to have played a major role in achieving these benefits.
> (Bell Island Commuters' Committee, 1983, p. 3)

In addition to making requests for assistance with promoting employment options on the island, the committee characterizes "the Bell Island Transportation System as an investment to aid in the development of a presently depressed economy" (p. 3). This investment is just, the committee argues, given the harsh mobilities regime they experienced after mining shut down:

> For many years our people had contributed to the cultural and economic life of our province. For a period after the closure of our iron ore mines, unofficially at least, government policy was "resettlement" ... "incentive grants" to depopulate the community ... followed by a period of "no policy" for Bell Island. Our people had been stigmatized as "dole recipients". The only solution seemed to be "the welfare net".
> (Bell Island Commuters' Committee, 1983, p. 3)

What is needed, instead, is for the government to "sign a 'Social Compact' with our people to ensure that service to our people will be the guiding principle in your deliberations" (p. 3). Such a compact would ideally lead to a new provider of the ferry service and an arrangement whereby "[o]ur people would have an effective voice in the operation of the service ensuring its responsiveness to their interests and needs. It would be unquestionably *our* Ferry System" (p. 4, emphasis in original). This inclusion of local needs, interests and visions could be achieved by hiring more Bell Islanders through a "local preference policy" (p. 6) and "*the implementation of the Standard User Committee into the operation of the ferry service* [...] would ensure effective responsiveness to the interests and needs of the community" (pp. 5–6, emphasis in original). Such changes would reflect a needed recognition by the government that "[a] most essential factor in Bell Island's existence as a people and survival as a community is the transportation system" (p. 1). For "[o]ur ferry service is inextricably tied to the very life of our community. For there to be a social and economic future for our town we must have the best, the most efficient transportation system possible" (p. 2).

By defending workers' right to daily commuting mobilities, the committee implicitly defends all island spaces in the province and the equal importance of aquamobilities to other forms of transportation. They are involved in defending and producing 'counter' mobilities and spaces in two senses. Henri Lefebvre (1991, p. 349) discusses how resistance can "inaugurate" a "counter-space" or "alternative to actually existing 'real' space". Bell Islanders' commentaries about their ferry service have countered political discourses that denaturalize the longstanding centrality of *both* aquamobility and nonurban residency in Newfoundland and Labrador. These efforts from the 1960s to 1980s laid the groundwork for continuing struggles for mobility justice by Bell Islanders and other residents of Newfoundland and Labrador. Through their ferry committees and other means, Bell Islanders have continued consistently to monitor and speak out about their ferry service.

The original central argument for mobility justice, developed by commuting workers at a time of alarming local unemployment, has continued. Ferry committees have provided a structure for continuing citizen-based local advocacy with respect to a core public service. On Bell Island, the need to defend this right is no less acute in the 2010s than it was in the 1960s and 1970s. As a former committee chair explained to me,

> It is a very difficult position to have, because really you're fighting for change that you have no power to make. [...] [W]e're not operational and we're not even governance ... we are ... [a] group of volunteers basically put in that position by the general public ... [O]ur role is to advocate for the people and to advocate for workers.
>
> (Semi-structured interview, 2016)

Despite these challenges, the committee has maintained an ongoing dialogue with ferry users and government officials, thereby serving as an information conduit between these two groups. They regularly request clarification and fuller details about the status of the service from government officials, and then they share that information with users through various means, including phone calls, meetings and social media. They have also continued to be the main body that requests improvements to the service or explains the potential impacts of changes to politicians, bureaucrats and media sources.

The concerns about schedules, capacity, rates and the state of the ferries assigned to the Bell Island service that were robustly articulated several decades ago remain acute today. However, the committee also registers new matters that arise. For example, last year a provincial policy was enforced that requires ferry users to leave their vehicles during the 20-minute crossing. This regulation has created difficulties for numerous individuals who have mobility challenges and for those who are in fragile health, such as oncology patients returning from chemotherapy treatments in St. John's. As in the past, the committee's commentaries have moved into the public sphere through media coverage:

> The chairman of the Bell Island Ferry Users Committee [...] fired off a detailed email to everyone from the provincial and federal human rights commissions to Prime Minister Justin Trudeau outlining his and his ferry users' long and ongoing frustrations with the service—particularly the policy of passengers having to leave their vehicles during the 15- to 20-minute ferry ride. ... "Our position as a committee, and the reason for asking for the risk assessment to occur, was due to our fears that the enactment of the policy of vacating vehicles would put our users with disabilities at risk."
>
> (Whiffen, 2018)

The Coalition of Persons with Disabilities Newfoundland and Labrador provided strong support for a risk assessment. As of spring 2018, Bell Islanders awaited its results. Their united calls for a reversal of this policy bolster abiding concerns for the mobility dignity of ferry users.

Conclusion

The work of Bell Island ferry committees has reinforced the notion that achieving mobility justice requires that residents of various localities have sufficient access to mobility infrastructures. These infrastructures are often sole conduits for securing livelihoods and accessing basic social services, thereby ensuring that residents from specific locations do not face the kind of social exclusion that can result from mobility injustice. This chapter contributes to literature on 'the right to the city' by highlighting the role played by rural residents in a voluntary grassroots organization to improve their community's access to basic public transit infrastructure and urban spaces. As in other commuting contexts, their commentaries have had a constitutive role in shaping more extensive and socially inclusive mobilities (Bissell, 2015). This example foregrounds the valuable contributions made by individuals and grassroots groups in seeking and safeguarding the conditions that allow for mobility justice. In Bell Island, as elsewhere, voluntary transit advocates are providing essential, but often invisible and unpaid, labor that allows for the effective development, modification or operation of public mobility infrastructures (Roseman, 1996). It is important to recognize the time, effort and intentionality that groups such as the Bell Island Ferry Users Committee devote to strategically crafted commentaries that push for such results.

As Mimi Sheller (2008, p. 31) notes, the "wide range of mobility struggles that have been ongoing over many centuries … [are] concerned with who has access to mobility, whose mobility is controlled by whom and who has a say in determining mobility systems and rights". Henri Lefebvre (1991, p. 365) cautions users of public spaces, who may want a voice in determining such systems and rights, against being "silence[d]" by the "abstract" power of bureaucracies and capital (see also Shields, 2013, p. 346). In parallel with urban solidarity movements that have appeared under the rubric of the right to the city, rural-based citizens' groups in many parts of the world defend and reappropriate their spaces and their mobilities (Roseman, 2008). The powerful language of ferry justice for Bell Islanders, which emerged in the wake of the 1960s industrial closure and the organization of the original Bell Island Commuters' Committee, continues to be deployed today.

The committee's original central argument was for mobility justice for unemployed working-class residents. Their call for a citizen-based say over a core public service explicitly opposed a mobilities regime that was forcing many residents into poverty or migration. In dialogue with Lefebvre's idea of the right to the city, I argue that rural-based citizens' activism, such as that dealing with transportation, should be considered alongside debates about the right of urban dwellers, especially those subordinated by classist, racialized and disabling public environments (Chouinard, 2001; Parks, 2016; Sawchuk, 2014). Since the 1960s, members of the Bell Island ferry committees and other Bell Islanders have consistently employed strategic commentaries to articulate why Bell Islanders have the right to stay in their home community, to access paid employment and basic social services and to move under dignified mobility conditions. They persist in registering, when necessary, their concerns about schedules, capacity, rates and the state of the ferries assigned to Bell Island.

Bell Island ferry committee members have played an indispensable and complex role since the 1970s. This complexity is particularly evident at times when there are periods of conflict or crisis with respect to the service. By maintaining an ongoing dialogue with government officials to suggest changes to the service, demanding full explanations about problems or changes and sharing the answers they receive, Bell Island ferry advocacy committees have reinforced, articulated and defended Bell Islanders' claim to a right to their island and their ferry system even in the most difficult periods.

Notes

1 This chapter is part of a larger study of the Bell Island ferry system conducted as part of On the Move: Employment-Related Geographical Mobility in the Canadian Context, a project of the SafetyNet Centre for Occupational Health & Safety

Research at Memorial University. On the Move is supported by the Social Sciences and Humanities Research Council of Canada through its Partnership Grants funding opportunity (Appl ID 895-2011-1019), Innovation NL, the Canada Foundation for Innovation, and numerous university and community partners in Canada and elsewhere. Warm thanks to Bell Islanders for their insights, and to Diane Royal and Lesley Butler for research assistance.

2 The provincial government has operated this service with unionized crew since 1989. It is now a two-boat service. MV *Flanders'* maximum capacity is 36 cars and 240 passengers. MV *Legionnaire*, which recently replaced MV *Beaumont Hamel*, can carry a maximum of 64 cars and 200 passengers.

3 TTC is the acronym for the Toronto Transit Commission.

4 Small boat ferries were central to the mining industry, with many miners commuting weekly between their homes in other communities and Bell Island.

5 When the government makes it available, resettlement can still occur if 90% of residents vote to leave a specific location in return for financial compensation (Barry, 2016).

6 Federal funding for some intra-provincial ferry routes, including Bell Island's, was provided from 1949 when Newfoundland joined Canada until the mid-1980s (CNM Inc., 1981; Collier, 2010).

7 The politician Steve Neary and others unsuccessfully advocated for an undersea tunnel to be built between Bell Island and Newfoundland.

8 The most recent election was held on October 1, 2017. The committee of 11 includes the mayor and another representative of the town council of Wabana; two commuter representatives; one representative from the business community; one youth representative; and two from the general public. The non-voting members are the Member of the provincial House of Assembly; the Regional Manager of Marine Services for the province; and a consultant to the Town of Wabana.

9 This meeting followed a report (CNM Inc., 1981) that was prepared in advance of the 1984 repeal of the 1954 provincial Ferries Act.

10 The gender-specific language reflects the dominance of androcentrism in discussions about commuting in this period, including among grassroots groups. Both women and men were part of the group of individuals who began to commute daily to St. John's in this time period.

References

Atlantic Development Board. (1967). 'Special housing assistance for Bell Island'. Press release.

Barry, G. (2016, November 29). 'N.L. government holds line on 90% threshold for resettlement'. *CBC News.* Retrieved from www.cbc.ca/news/canada/newfoundland-labra dor/resettlement-policy-review-1.3873176.

Batterbury, S. (2003). 'Environmental activism and social networks: Campaigning for bicycles and alternative transport in west London'. *The Annals of the American Academy of Political & Social Science*, 590(1), 150–169.

Bell Island Commuters' Committee. (1983, October 26). 'A brief presented to the Newfoundland Provincial Cabinet by The Bell Island Commuters' Committee (on behalf of the people of Bell Island) regarding "The Bell Island Ferry Service"'. Bell Island, Newfoundland.

Bell Island Commuters' Committee and Representative Community Groups. (1973). 'A brief regarding the Bell Island ferry system'. Submitted to the provincial premier Frank D. Moores. Bell Island, Newfoundland.

Bissell, D. (2015). 'How environments speak: Everyday mobilities, impersonal speech and the geographies of commentary'. *Social & Cultural Geography*, 16(2), 146–164.

Butler, C. (2012). *Henri Lefebvre: Spatial politics, everyday life and the right to the city.* London: Routledge.

Chouinard, V. (2001). 'Legal peripheries: Struggles over disabled Canadians' places in law, society and space'. *The Canadian Geographer*, 45(1), 187–192.

CNM Inc., (1981). *Appraisal of Newfoundland's intra-provincial ferry services. Volumes I & II.* Prepared for the Government of Newfoundland and Labrador. St. John's, NL: CNM Inc.

Collier, K. (2010). 'Bridging the gulf: Coastal and ferry boats after confederation'. *Heritage Newfoundland and Labrador.* Retrieved from www.heritage.nf.ca/articles/economy/boats-after-confederation.php.

Cook, N. & Butz, D. (2016). 'Mobility justice in the context of disaster'. *Mobilities*, 11 (3), 400–419.

Cook, N. & Butz, D. (2010). 'Narratives of accessibility and social change in Shimshal, northern Pakistan'. *Mountain Research & Development*, 31(1), 27–34.

Cresswell, T. (2010). 'Toward a politics of mobility'. *Environment & Planning D: Society & Space*, 28(1), 17–31.

Gendreau, R. (1966, May 24). 'What are they doing for island's future'. *The Daily News*, p. 4.

Glick-Schiller, N. & Salazar, N. (2013). 'Regimes of mobility across the globe'. *Journal of Ethnic & Migration Studies*, 39(2), 183–200.

Guy, R. (1966, December 3). 'Unemployment protection fund urged for one-industry towns'. *The Evening Telegram*, p. 3.

Harvey, D. (2008). 'The right to the city'. *New Left Review*, 53, 23–40.

Hine, J. (2011). 'Mobility and transport disadvantage'. In M. Grieco & J. Urry (Eds.), *Mobilities: New perspectives on transport and society* (pp. 21–39). London: Ashgate.

House of Commons Debates. (1966). *First session, twenty seventh parliament, volume X.* Ottawa: Roger Duhamel. Retrieved from http://parl.canadiana.ca/.

Huang, J. (2014). 'A fair deal for transit riders'. *Our Times*, 33(2), 16–19, 21.

Hulchanski, J. D. (2010). *The three cities within Toronto: Income polarization among Toronto's neighborhoods, 1970–2005.* Toronto: Cities Centre.

Ilcan, S. (2013). 'Introduction: Mobilities, knowledge, and social justice'. In S. Ilcan (Ed.), *Mobilities, knowledge and social justice* (pp. 3–23). Montreal, QC: McGill-Queen's University Press.

Kesselring, S. (2014). 'Mobility, power and the emerging new mobilities regimes'. *Sociologica*, 1, 1–30.

Lefebvre, H. (2003[1970]). *The urban revolution* (R. Bononno, Trans.). Minneapolis, MN: University of Minnesota Press.

Lefebvre, H. (1995). 'The right to the city'. In E. Kofman (Ed.), *Henri Lefebvre: Writings on cities* (pp. 147–159). New York: Blackwell.

Lefebvre, H. (1991). *The production of space* (D. Nicholson-Smith, Trans.). Oxford: Blackwell.

Lefebvre, H. (1968). *Le droit à la ville* (2nd edition). Paris: Anthropos.

Martin, M. (2017[2006]). 'The resettlement program'. *Heritage: Newfoundland and Labrador.* Retrieved from www.heritage.nf.ca/articles/politics/resettlement-program.php.

McDonagh, J. (2006). 'Transport policy instruments and transport-related social exclusion in rural Republic of Ireland'. *Journal of Transport Geography*, 14(5), 355–366.

Parks, V. (2016). 'Rosa Parks redux: Racial mobility projects on the journey to work'. *Annals of the American Association of Geographers*, 106(2), 292–299.

Purcell, M. (2014). 'Possible worlds: Henri Lefebvre and the right to the city'. *Journal of Urban Affairs*, 36(1), 141–154.

Purcell, M. (2002). 'Excavating Lefebvre: The right to the city and its urban politics of the inhabitant'. *Geojournal*, 58(2–3), 99–108.

Roseman, S. (2013). 'Unemployment and labor migration in rural Galicia (Spain)'. *Dialectical Anthropology*, 37(3), 401–421.

Roseman, S. (2008). *O Santiaguiño de Carreira: O rexurdimento dunha base rural no concello de Zas*. A Coruña: Baía Edicións.

Roseman, S. (1996). '"How we built the road": The politics of memory in rural Galicia'. *American Ethnologist*, 23(4), 836–860.

Sawchuk, K. (2014). 'Impaired'. In P. Adey (Ed.), *The Routledge handbook of mobilities* (pp. 409–420). London: Routledge.

Schwartz, J. (2014). 'Classrooms of spatial justice: Counter-spaces and young men of color in a GED program'. *Adult Education Quarterly*, 64(2), 110–127.

Sheller, M. (2011). 'Sustainable mobility and mobility justice: Towards a twin transition'. In M. Grieco & J. Urry (Eds.), *Mobilities: New perspectives on transport and society* (pp. 289–304). London: Ashgate.

Sheller, M. (2008). 'Mobility, freedom and public space'. In S. Bergmann & T. Sager (Eds.), *The ethics of mobilities: Rethinking place, exclusion, freedom and environment* (pp. 25–38). London: Routledge.

Shields, R. (2013). 'Lefebvre and the right to the open city?' *Space & Culture*, 16(3), 345–348.

Staeheli, L., Dowler, L. & Wastl-Walter, D. (Eds.). (2002). 'Social transformation, citizenship and the right to the city'. Special issue of *Geojournal*, 58(2–3).

The Daily News. (1973a, August 2). 'Commuters to meet premier'. *The Daily News*, p. 2.

The Daily News. (1973b, August 31). 'Extra trips for ferry'. *The Daily News*, p. 18.

The Daily News. (1972a, August 1). 'Residents get extra ferry trip across the tickle'. *The Daily News*, p. 3.

The Daily News. (1972b, August 11). 'CTC investigation planned for Island ferry service'. *The Daily News*, p. 3.

The Daily News. (1966a, May 16). 'Miners will telegram Pearson and O'Keefe'. *The Daily News*, p. 3.

The Daily News. (1966b, May 30). 'Present ferry petition to JRS today or Tuesday'. *The Daily News*, p. 3.

The Daily News. (1966c, June 1). 'Wants free ferry'. *The Daily News*, p. 16.

The Daily News. (1960, September 15). 'Committee interviews Premier'. *The Daily News*, p. 2.

The Evening Telegram. (1966a, December 1). 'Committee investigating Bell Island's problems'. *The Evening Telegram*, p. 3.

The Evening Telegram. (1966b, December 5). 'New rates on ferries'. *The Evening Telegram*, p. 4.

Vannini, P. (2012). *Ferry tales: Mobility, place and time on Canada's west coast*. London: Routledge.

Vannini, P. (2011). 'Mind the gap: The tempo rubato of dwelling in lineups'. *Mobilities*, 6(2), 273–299.

Velaga, N., Beecroft, M., Nelson, J., Corsar, D. & Edwards, P. (2012). 'Transport poverty meets the digital divide: Accessibility and connectivity in rural communities'. *Journal of Transport Geography*, 21, 102–112.

Weir, G. (2006). *The miners of Wabana: The story of the iron ore miners of Bell Island.* St. John's, NL: Breakwater Books.

Whiffen, G. (2018, February 16). 'Bell Island ferry users group steps up its fight against policy that passengers must leave vehicles'. *The Telegram*. Retrieved from www.thetelegram.com/news/local/bell-island-ferry-users-group-steps-up-its-fight-aga inst-policy-that-passengers-must-leave-vehicles-186578.

Yeoman, E. & Roseman, S. (Dirs.). (2016). *Honk if you want me off the road.* Documentary Film. St. John's, NL: Humanities and Social Sciences Film Unit, Memorial University. Retrieved from https://vimeo.com/196130980.

Justice and biomobilities

11 *Black As*

Performing Indigenous difference

Georgine Clarsen

Introduction

In 2014, four Yolngu men from the Arnhem Land township of Ramingining in Australia's north emailed a video clip to David Batty, who has been making films with Indigenous people for over 30 years.[1] Chico, Jerome, Dino and their adopted white brother, Joe, sent Batty footage of their crocodile hunting exploits and invited him to create a television series with them. The result was the 24-part series *Black As*. Shot during the 2015 Dry season in the Djambarrpuyngu language (of the Yolngu Matha language chain) with English subtitles, this low-budget series, of which the four stars are co-owners, tells a comic story of a hunting trip.[2] The series went directly to web streaming via the Australian Broadcasting Corporation, and quickly became one of the broadcaster's highest-rated Iview programs, with more than 1.2 million views nationally for full episodes and 23 million hits worldwide on one promotional clip alone. Batty's media company, Rebel Films, has so far attracted some 150 million viewers around the world, including for the Spanish-language version on Facebook and YouTube. As I write, Batty and his crew are in post-production on the second *Black As* series, financed by ABC Iview sales and a crowdfunding campaign.

In this chapter, I analyze this lighthearted representation of young Indigenous men's movement across their Country as an enactment of mobility justice. Following Tim Cresswell's (2010) formulation of mobility as a political entanglement of movement, representation and embodied practice, I read *Black As* as a specifically Indigenous expression of mobility, which is necessarily politicized in the face of the ongoing settler-colonialism oppressions that have imposed deep inequalities and uneven mobilities on Indigenous subjects. Exemplifying what Audra Simpson (2014) calls a politics of refusal, the men move their unreliable vehicles across the unsmoothed terrain of their Country. They deploy their embodiment as encultured Indigenous men of that place to move on their own terms, outside settler movement regimes, to find bush foods and tell their story to camera. They thereby reanimate (as filmic representations) for a new generation and circumstances cultural practices that are tens of thousands of years old. As Simpson demonstrates, Indigenous refusal is more than resistance to injustice; it is also a political, generative, embodied assertion of ongoing sovereignty. In this chapter I tease out the story presented in *Black As* as a refusal of the historical and contemporary forces that have sought to constrain Indigenous mobilities and deny Indigenous sovereignty. Dino Wanybarrnga, Chico Wanybarrnga, Jerome Lilypiyana and Joe Smith's performances of mobilities and (essential to the humor of the series) immobilities offer a proud story of self-determined movement, which they have mobilized across their immediate community and the globe with the help of non-Indigenous filmmaker allies.

Social justice, mobility justice

Few non-Indigenous Australians have first-hand experience of life in townships like Ramingining, which is located some 600 kilometers east of Darwin on the edge of the Arafura Swamplands, accessible only by an expensive air service or roads that are impassable during the wet season. Most, however, are familiar, through media reports, with the indicators of injustice in Indigenous communities: high levels of poverty, accusations of endemic dysfunction, shocking health and educational statistics and high suicide and incarceration rates, especially for young men (Chenhall & Senior, 2017). As a prominent commentator on Australia's north put it recently, the general picture promulgated in non-Indigenous Australia is of "restless, poorly schooled, vulnerable teenagers brought up in pulverized communities in the bush" (Rothwell, 2017, p. 16).

These four Yolngu men, however, employ a growing tradition of cross-cultural regimes of representation to tell a different story about their lives in new times – a positive and inspirational account of their life in two worlds. As they travel across their Country, the men record a story of continuing mobility in which their desire and capacity for movement is not overdetermined by the injustices and inequalities that surround them (Sheller, 2016). Their story, molded to their own circumstances and framed by Indigenous protocols that remain deeply rooted in the landscape, is one of going places by being in place.

For these men, the swamp, rivers and sea terrain surrounding Ramingining is *their* place. In Aboriginal English, it is their Country or – as each of them is differently connected to particular locations – Countries (Rose, 2017). The performances they record reference a history that stretches back at least 50,000 years. These Yolngu men's everyday sense of connection to the thousands of generations of ancestors who have lived in that place and brought it into being is central to the action, but it is lightly conveyed as a reality that does not need explaining. Doreen Massey (2005, p. 9) was not thinking of the Australian *longue durée* when she wrote of places brought into being by interrelations that are both temporal *and* spatial, but her arguments provide a productive framework for understanding the spaces represented in *Black As* as the dynamic, relational achievement of both human and more-than-human actors, where "heterogeneity coexists" across space and time. As episodes unfold, the men's story reveals something of the processes that keep Country alive, cared for and activated in the present. Country, in this sense, may be understood as an array of deeply known locations and resources that have been created and maintained by the movement of the 'old people' – thousands of generations of Ancestors, both human and more-than-human – across it, right into the present. So the everyday mobilities and immobilities recorded in *Black As* are specific to that place and its deep history. Those (im) mobilities confer contemporary meaning to that place through the cultural protocols, law and knowledge these men practice as they move across it, thereby reinscribing Indigenous sovereignty.

With minimal directorial control or scripting, Batty and his crew move with the men across unsmoothed terrain that constantly impedes their mobility. They film each episode in sequence, in local language and on actual locations, with the script and action unfolding according to the storytelling traditions long practiced in Indigenous cultures (Langton, 2003, p. 109). Finding bush food – fish, eggs, birds, shellfish, turtle, crabs, wallaby and crocodile – is the men's obsession, sustenance and pleasure. Their hunting skills with a wooden harpoon and fish hook are integral to their status as knowledgeable, entitled, cultured men of that place. So too is their anarchistic incorporation of modern modes of mobility across land and water. The rough conditions, as well as Ramingining's isolation and poverty, impel these men to make their own mobility however they can. They cobble movement out of whatever they can lay their hands on, be it wrecked motorcars rescued from the bush, a dinghy with an unreliable outboard motor that was "humbugged" from an uncle or canoes that were quickly shaped with an axe from sheets of discarded corrugated iron. As they do so, they create a proud, touching, funny, anarchistic and important story of *both* cultural continuity *and* cultural transformation, of survival and change.

These young men are part of a growing cohort that, in a society traditionally controlled by elders, is ambitious to tell their stories. They exemplify what Marcia Langton (2003, p. 110) characterizes as "a new permissiveness, atypical of the old traditions". Increasingly, as Melinda Hinkson (2005) observes of Warlpiri engagements with new media in the Central Desert, this is a cohort that is well positioned to embrace the opportunities for intercultural alliances presented by digital technologies. The Ramingining men have taken the initiative to claim the status of cultural brokers, empowered to create a story about their lives that they direct to a local, national and global audience. Their story is set to a Ramingining version of 'crossover music' played by local musicians. It draws on both *Yidaki* (didgeridoo), as well as click sticks and traditional vocals, and newer traditions (blues, reggae, country, pop and surfing riffs played on guitars, drum kit and keyboards) that have become the music of everyday community life. For viewers, the distinctive and buoyant soundtrack serves to emphasize the incorporation of the non-Aboriginal world into an Aboriginal cosmology. Through such choices, the men affirm and celebrate their cultural distinctiveness as Yolngu and "lessen the pressure for Aboriginal people to become incorporated or assimilated into the global worldview" (Langton, 2003, p. 110).

The brothers' performances speak of being deeply rooted in their place, of moving through it with an historically nuanced sensibility that cannot simply be translated into Western conceptual frameworks. Their slapstick stories of thwarted mobility – always overcome through muscle power, local knowledge, fractious cooperation and technological smarts – reveal the men mobilizing themselves across both space and time, resisting colonial mobility regimes to re-enact sovereignty. As viewers, we celebrate their triumphs, their off-the-wall ingenuity and their Yolngu way of being together in the world, at the same time as we apprehend some of the forces that work toward their containment and immobilization in a settler-colonial polity. On the periphery of the action we glimpse some of the realities in which Yolngu live under the Northern Territory Emergency Response, a regime of heightened federal government surveillance and control commonly known as 'the Intervention' (Rothwell, 2017; Scott & Heiss, 2015). These men's stories of a contemporary tradition of living on their land – funded, recorded, viewed and distributed through the power of the latest digital technologies – provide a unique perspective on questions of mobility justice that are simultaneously past and present, and played out in material, imagined and virtual realms.

Black As: the story

At the close of *Ten Canoes*, the feature film that first brought Ramingining and the Arafura Swamp to an international audience, the actor David Gulpilil pronounced, "Now you've seen my story. It's a good story. Not like your story, but a good story all the same" (de Heer & Djigirr, 2006). *Black As* is a good Ramingining story too. It is nothing like the stories of crime and dysfunction usually told about young Aboriginal men in Australia, stories that were used to justify the Intervention and through which Indigenous men's lives are increasingly criminalized (Anthony, 2017b). Nor is it an epic morality tale with the ambitions and budget of a *Ten Canoes*. However, it is part of that "minor social revolution" of Indigenous involvement in film, television and now digital products that foregrounds survival and expresses Indigenous preoccupations with Country and culture on Indigenous terms, in Indigenous ways (Langton, 2003, p. 113; Rothwell, 2006).

The *Black As* storyline is simple and shares many elements with David Batty and Francis Jupurrula Kelly's earlier *Bush Mechanics* series (Batty & Kelly, 2001; Clarsen, 2002; Paul & Bolognese, 2017). The Wet season is over in the "Wilds of Arnhem Land", as the publicity material puts it, and the people of Ramingining are sick of expensive store food. With the bush tracks drying out, four brothers decide to "travel their ancestral lands in search of adventure and a good feed". They locate their wrecked car – an early model Suzuki Vitara – crushed under trees that were uprooted a few months earlier by cyclone Lam. Using only an axe and a wheel brace, they get their Suzi running again and drive the red dirt track into town. Encouraging each other, the men nervously go to an uncle's house to borrow his boat (a precious possession in Ramingining) for a couple of days. "We'll fill your empty freezer with turtle meat", they promise. Oppy, growling and reluctant, has heard this before. "Your last chance", he tells them. They drive out of town happy and minimally equipped for camping: "Yolngu style, not white way". Foreshadowing the mishaps to come, the boat bounces off the trailer soon after they leave town. The men promise to help each other so they won't get into trouble with Oppy, but before the boat is launched into the river, Joe sinks their car and trailer in the tidal mud. They abandon the bogged car and trailer, laboriously push the boat into the river and head to the open sea at full throttle and with high spirits. They camp at Moororonga Island, part of the Crocodile Island group at the very top of the continent, to fish and hunt for the turtle meat they promised Oppy.

And so the episodes unfold. There is abundant wild food, stunning landscapes, exhilarating fishing scenes, comic misadventures, good-humored chiacking and, occasionally, more heated recriminations over who was responsible for particular disasters. These men are beautiful, and their portrayal is a familiar narrative of young masculine friendship. They wrestle, play, tease and trick each other. Their humor is crude. True to their age, their days are punctuated by hunger and sleep. But before too long the wheels are falling off their camping adventure, and things become serious. Jerome has forgotten to bring fresh drinking water, and they must leave the idyllic island in a hurry, without their prized turtle, that escaped overnight. The outboard motor fails on the open sea, and, overcome with heat, they are reduced to drinking Jerome's piss out of a whiskey bottle found discarded on the beach,

Figure 11.1 Sunk in Tidal Creek

with which he had earlier tricked them. With the men expiring from dehydration, the boat drifts back into the mangrove swamps somewhere on the mainland. They walk, lost, across baking salt pans, Dino dragging two small crocodiles for food. The men are saved by finding water stored in the trunk of a paperbark tree, which Jerome releases with their ubiquitous axe. Restored, they find a billabong, where they swim and settle down to a big feed of crocodile meat.

The lessons in survival are many: good tricks like finding bush food, making a harpoon from a slender length of timber, locating drinking water in a tree or starting a fire with a shotgun cartridge and an electric toothbrush. But the mood has shifted gear. Mythic and darker elements, hinted at earlier in the story, become palpable. A swirling dream sequence, represented by traditional bark paintings of turtles and crocodiles, suggest that ancient and powerful forces surround them. They observe protocols about whose Country they are entering and who should walk first to announce their intrusion. ("I'll go first, you don't know this place. You'll make this place disappear".) They sense bad spirits around them, the *Mokuy'guun*, and they sleep close together holding their harpoon for reassurance.

Those spiritual references reveal the continuity between the settler colonial present and the Indigenous *longue durée*. They demonstrate the gulf between this series and the reality TV show *Survivor*. This is not a battle against hostile nature in a wild location. These men occupy a thoroughly encultured landscape, which they experience as redolent with life, both past and present. They playfully allude to those connections in the episode called "Tin Canoes", in which they cross a croc-infested billabong in canoes shaped from sheets of rusty iron. The scene mimics the ten canoeists poling across the Arafura Swamp, preserved in the iconic Thomson Time photographs of 1937, when life in Ramingining was barely impacted by invasion, as well as the subsequent period of cultural renewal portrayed in *Ten Canoes* (2006).[3] The men's decidedly cheeky reference to those images is a conscious intertextual joke that they know their whitefella audience will recognize. At the same time it suggests the "new forms of Indigenous cultural memory" that such filmic representations of Ramingining have generated, although these are more youthful, irreverent and hybrid representations of Yolngu identity than in *Ten Canoes*, made just ten years earlier (Davis, 2006).

And so the men draw their story to its comic conclusion. At Mangbirri, an eerily deserted outstation, they repair an abandoned Hyundai and, after many frustrating misadventures, use it to haul their Suzuki and trailer out of the tidal mud. Unable to find Oppy's boat, which is lost somewhere on the river, they conclude that "bad spirits" have hidden it. The boat is finally located and, with much exertion, wrangled onto the trailer. After more mishaps, with the boat and trailer bouncing precariously behind their Hyundai, the four speed back into Ramingining and push the damaged boat back into Oppy's yard. They are a week late, without turtle meat, his cooler box filled with rotting fish. Oppy, predictably angry, shouts, "Never again you bastards!" The

Figure 11.2 Mangbirri: choosing a new car

men beat a hasty retreat in their car, after presenting him with their bottle of 'whiskey' and inviting him to "drink up and forget about your boat". Just down the road they spy another boat, newly arrived from Darwin and much better than Oppy's. Their optimism and camaraderie are instantly restored: "Imagine us boys shooting along in that! Let's borrow it and go back for turtle. Can't hurt to ask!" To the sound of a tropical thunderclap they make their move, poised to humbug this new, unsuspecting uncle.

Living in Ramingining

Central to the power of the series is the challenge it poses to settler assumptions about life in remote Aboriginal communities. Far from sedentary, poorly schooled and vulnerable, *Black As* represents men who are mobile, resourceful and full of storytelling talent. They skillfully meld customary practices and narrative modes with contemporary technologies of mobility and communication into an engaging vehicle for expressing, affirming, celebrating and disseminating what is important to them. As Batty declared in his Pozible campaign pitch,

> These guys want to tell the world and their mob that there is plenty to celebrate in Aboriginal communities. Living off the land and being on Country is not only fun, it's much healthier than relying totally on store

food and sitting around doing nothing. These guys want to inspire others to do the same.[4]

Such recuperative ambitions, communicated via contemporary media and addressed to both local audiences and outsiders, give this series a "unique in-betweenness", to use Aboriginal filmmaker Darlene Johnston's term (Davis, 2011). The sense of complex entanglement that saturates the series is visually highlighted (but entirely unremarked) by the fact that one black brother is white.

After one of their big "feeds", the four friends discuss how they will educate their children. "Will they grow up knowing culture, like us?" they ask. Their children, they agree, should be able to go out bush and not be scared. They will need to learn to live with the land as well as go to school. Mostly, however, their message of cultural affirmation is simply part of the story, represented without comment or special emphasis. They communicate their sheer joy of being on Country, for example, in the second-last episode they called "Spring Clean" when the men are return-ing to Ramingining. Cranky, covered in mud and stinking, they visit a beautiful freshwater spring to drink clean water and take a "proper wash". Their good humor returns as they walk into this especially magical place: "Present yourself to the country", they tell each other in mixed Djam-barrpuyngu and English. "Like I'm a stranger? Yes, or the water will turn salty. Can't you talk for me? This is our grandmother's place, Joe". Joe responds: "I'm introducing myself. I'm following these Yolngu boys", he tells the Ancestors. "Say 'it belongs to our grandmothers'", he is instruc-ted. "It's their grandmothers", he repeats to establish his credentials. They drink with pleasure and an air of reverence. Joe tells his friends, "This is called 'spring water' in English". There follows a joyous sequence of boyish water play, of washing bodies and clothes. As they leave the spring, Joe offers one last acknowledgement to the Ancestors: "Thank you old people for the water".

"Suzi Send-Off", an episode in which they cremate their much-loved Suzuki, similarly reveals their entanglement in both Yolgnu and whitefella worlds, and the complex processes whereby their car becomes Yolngu through its incorporation into cultural protocols. After they haul their wrecked "best car" out of the mud, they realize it is beyond repair. Although it is hard to imagine anything of value remains, they worry people will strip it. So the men tow the car to a quarry for a ritual farewell, "a proper smoking ceremony", rather than dump it where it was bogged. They sorrowfully fill the car with dry grass and green leaves, and sit together as it burns. Joe berates himself, "How could I sink it!" With heartfelt declarations, they comfort each other and recall the many good times they had in it.

I remember one time skidding along the mud. We drove it all the way from Darwin – 140! Foot down! The first night it stayed at my place. It took us to men's business many times. ... That's it. It's having a rest at last. Rest in peace. It's going to car heaven.[5]

They each step forward and, with traditional gestures of respect and mourning, throw a handful of sand on the blackened metal. Comic and touching at the same time, the scene portrays a thoroughly Yolngu automobile. The four collectively own the vehicle and, at great expense, had driven it into their community. It had been endlessly modified, with more and more parts falling off (or discarded) to keep it functioning in the rough terrain and through hard usage. The mobility their car has provided, the labor and inventiveness it embodies and their memories of good times in it are central to their identities as men who refuse settler attempts to erase Indigenous difference.

The car, however much loved, simultaneously exemplifies a mobility of poverty and deprivation characteristic of Indigenous townships, where new forms of settler governance are continually (re)imposed. As long as they can be made to go, rusted cars – with shredded tires, missing windscreens, no seatbelts (not to mention seats), doors tied shut with a piece of string – count as prized objects in this economy of scarcity and long distances, particularly for young men. Apart from white workers or some successful artists, few individuals in these communities are able to afford a late-model car. Those who can find their cars are soon destroyed by the obligation to share and the rough tracks. In fact, new cars with their computerized controls are a liability in remote townships where there are no specialist mechanical services. As in Aboriginal communities across the Northern Territory, the yard of the Ramingining Police Station holds a number of late-model cars in perfect order – except for a missing transponder key (electronic immobilizer), which is essential to start the engine and unlock the steering. Even a simple fault like a jammed retractor on a seat belt can render a late-model vehicle illegal. The cost of returning cars to the manufacturer is so prohibitive that such vehicles are effectively worthless. Aging cars, built before electronic ignition and fuel injection systems became standard – like the men's Suzuki and Hyundai – that can be repaired with bush-mechanic skills are an appropriate and affordable technology for remote settings, although they carry particular risks. Indigenous people in the Northern Territory, who routinely drive long distances on dirt tracks, are almost three times more likely than other road users to die or sustain serious injury in a road accident (Anthony & Blagg, 2013, p. 52; Dempsey, 2017). It is precisely those bush cars – unroadworthy, dangerous and unregistered – that the police target as part of the normalizing imperative of the Northern Territory Intervention.

The Intervention was imposed in 2007 by a conservative federal government and retained by subsequent governments as an 'emergency response' to allegations of endemic child sexual abuse, violence and neglect. It remains in force, rebadged as the Stronger Futures Policy. The federal government's overriding of Northern Territory powers, which initially involved sending troops into 73 remote communities and required the suspension of the Racial Discrimination Act, remains contentious within the affected townships and beyond (Rothwell, 2017). Many senior women and some men from inside and outside those communities have expressed support for its positive impact on the lives of women and children, particularly its emphasis on greater policing, health checks for children, curfews, restrictions on alcohol and gambling and the quarantining of welfare income via a BasicsCard to buy food and other essentials. Others, including senior men and women in Ramingining, have condemned the Intervention for its lack of consultation, the dismantling of community employment programs (CDEP), the takeover of lands, the suspension of community councils, the ban on bilingual education, mandatory business plans for community organizations, the increased use of white contractors and the return to presumptions that Aboriginal people should be assimilated into settler institutions (Altman, 2016; Scott & Heiss, 2015).

Notwithstanding these divergent views, the Intervention has worked to demonize Aboriginal men in the Northern Territory. Men have expressed anger, pain and humiliation that they all should be stigmatized and disempowered whether or not they are engaged in child abuse, violence, grogrunning or other destructive practices (Browning, 2011; Konishi, 2011). Certainly, the increased police presence has not uncovered the widespread 'paedophile rings' used to justify government action. Research, however, reveals that introducing federal policing in Aboriginal communities has coincided with a dramatic increase in the criminalization of Indigenous men, not for crimes of violence or child abuse, but for the relatively minor offenses of driving unregistered, unlicensed or uninsured. With the frequent nonpayment of fines, arrests have produced even higher incarceration rates than before the Intervention. In the Northern Territory, the imprisonment rate is increasing more rapidly than in any other jurisdiction in Australia, where Indigenous people already suffer the highest rates of any people on earth (Anthony, 2017a); although they comprise only 30% of the population, 85% of prisoners are Indigenous (Anthony, 2015).

The mobilities portrayed in *Black As* should be read within the broader context of the Intervention and its ramifications for Indigenous masculinities in remote communities. There state apparatuses place mobile men in an impossible situation. As soon as the brothers drive on public roads without a white film crew behind them, they risk being criminalized for a practice that the series demonstrates is a creative, localized response to the possibilities that modern technologies offer. The *Black As* series stands as a powerful expression of hope in the face of the disjuncture between their desire to keep Yolngu culture strong and in its place (in spite of the renewed colonizing imperative that the Intervention presents) and their desire to face outward to forge a life in a globalized world. From this perspective, these men are asserting their desire for, and right to, *both* mobility *and* immobility – virtual, imaginative and material. Their story demonstrates the "coexisting heterogeneity" (Massey, 2005, p. 9) of spaces, times and cultures, which are global and local and within which radical difference flourishes.

Conclusion: mobility justice and Indigenous difference

Yolngu difference and, indeed, superiority in a capitalist globalized world is humorously pitched in a series of mock commercials for local products, interwoven throughout the series and posted on their YouTube channel.[6] The advertisements were shot with Batty's vintage wind-up Bolex camera. A knob on the side of the camera allows him to change the speed while filming, giving the footage a Chaplinesque look. The actors added the soundtrack after the segment was edited. The Bolex gives a distinctive sensibility to their "Four Brothers Trading Company" advertisements, which promote local products that are freely available, without the need for the BasicsCard. Dino spruiks a method of filtering water through paper bark: "Crystal Clear^{TM}: When you want clean water, select the best. A fresh product from Swamp Enterprises". Two girls wash their hair in a spring – "Salon Pandanus – Hair for Today" – and harvest bush lollies – "Tamarind Bush Sweets^{TM}: Sweet As!" Chico makes paperbark sandals for children walking through the bush: "If you've got sensitive souls, think paperbark. Paperbark^{TM}: Tradition in Arnhem Land – Since Forever". Chico demonstrates an insect repellent made from mangrove mud: "MudroGard^R: 100% Natural. Tough as old dirt". And Joe, in the only acknowledgement of his fair skin, mixes water and ashes from the fire to create a sunscreen. Other advertisements ridicule bizarre practices of the capitalist world like cosmetic surgery. The "Qlay Clinic: A new Fashion for Faces (out past the bridge and turn left)" offers reconstructed noses, modeled from clay and grotesquely pointed in whitefella style.

These vignettes articulate a conspicuous disregard, even scorn, for the commodity value of capitalist products. Similarly, the men's cars are prized objects, no matter how battered, wrecked, worthless or illegal in the whitefella world. They are redefined and flexibly adapted to mitigate conditions in a remote community which lacks the personal wealth and collective infrastructure associated with car ownership in settler regimes. Here is exemplified the politics of refusal. In Simpson's (2014) terms, these Yolngu men refuse to stop being themselves, thereby asserting that colonization is incomplete. They insist on their emplaced identity in the face of settler certainty that their distinctiveness should be erased. This refusal does not mean, however, that the men have turned their backs on the world outside Ramingining. Far from it. Indigenous people in Australia have long been early adopters and creative reinscribers of new products brought to their Countries (Clarsen, 2017a, 2017b). Communities have embraced the power of mobile media, based on a history of early engagements with all forms of moving image, as they became available. Initially it was film and screen; then videocassettes, DVDs and television; and now digital technologies on laptops and mobile devices (Hinkson, 2005). Digital media, like all media, have changed the possibilities for cross-cultural communication, consultation and collaboration (Davis & Moreton, 2011; Taylor, 2012). They afford unprecedented opportunities for local low-budget production, new modes of viewing and exponentially accelerated opportunities for dissemination, with and without the physical mobility of people and things across space.

Black As, a seemingly simple story of masculine adventure, stands as a complex representation of mobility justice by insisting on dwelling in the wider world and one particular place at the same time, refusing to be immobilized and insisting on Indigenous sovereignty and ways of being in the world. It is a product of a constellation of forces: of the material, imaginative and virtual work of negotiating differences across time and space; of reworking Indigenous identities and geographies for a new generation; of affirming unceded sovereignty over lands and lives; and of raising questions of social justice played out through new forms of movement. The series stands as a timely reminder that mobility justice is a quality that inheres not simply in the capacity of individuals to freely move across generic space, which is usually celebrated as a right conferred by modernity. Rather, these modern men have produced a collective, communal assertion of mobility justice that reveals a specific conjunction of power and inequality. Together, these men from the particular place that is Ramingining are shaping their own terms of engagement with a hybrid world. In their world, Indigenous and settler cultures coexist, and new ways for them to move or stay – or both – are constantly opened up for renegotiation.

Notes

1 Yolngu is the name Aboriginal people in Arnhem Land use for themselves. I thank David Batty for our interview in August 2017 and his willingness to engage my questions since. Thanks also to members of the Menzies School of Health at Charles Darwin University, particularly Dr. Jenne Roberts and Dr. Dino Hodge. I also thank Lisa Slater and Trish Luker for helpful comments on earlier drafts.
2 The series and other *Black As* material is now available through Rebel films at http://rebelfilms.com.au.
3 Thomson Time refers to the late 1930s when Donald Thomson, white anthropologist and strong advocate for Yolngu independence, came to live with them. He recorded their lives in a series of glass plate photographs which inspired the film *Ten Canoes*.
4 See www.youtube.com/watch?v=y3MKx7ZmVbw.
5 See www.youtube.com/watch?v=id5gIDkWOi8.
6 See www.youtube.com/channel/UCtbwzLmInXpvtvmNoz_9wsg/featured.

References

Altman, J. (2016). 'Bawinanga and CDEP: The vibrant life, and near death, of a major Aboriginal corporation in Arnhem Land'. In K. Jordan (Ed.), *Better than welfare? Work and livelihoods for Indigenous Australians after CDEP* (pp. 175–218). Canberra, AU: ANU Press.

Anthony, T. (2017a). 'FactCheck Q&A: Are Indigenous Australians the most incarcerated people on earth?' *The Conversation*. Retrieved from https://theconversation.com/factcheck-qanda-are-indigenous-australians-the-most-incarcerated-people-on-earth-78528.

Anthony, T. (2017b). 'NTER took the children away'. *Arena Magazine*, 148, 21–25. Retrieved from https://arena.org.au/nter-took-the-children-away-by-thalia-anthony.

Anthony, T. (2015). 'Paperless arrests are a sure-fire trigger for more deaths in custody'. *The Conversation*. Retrieved from https://theconversation.com/paperless-arrests-are-a-sure-fire-trigger-for-more-deaths-in-custody-42328.

Anthony, T. & Blagg, H. (2013). 'STOP in the name of who's law? Driving and the regulation of contested space in Central Australia'. *Social & Legal Studies*, 22(1), 43–66.

Batty, D. & Kelly, F. (Dirs.). (2001). *Bush mechanics: The series* [Television series]. Retrieved from www.pawmedia.com.au/productions/bush-mechanics-4.

Browning, D. (2011, April 16). 'Elders on the intervention. ABC Radio National Awaye!' [Program podcast]. Retrieved from http://mpegmedia.abc.net.au/rn/podcast/2011/04/aye_20110416_1812.mp3.

Chenhall, R. & Senior, K. (2017). 'Living the social determinants of health: Assemblages in a remote Aboriginal community'. *Medical Anthropology Quarterly*. doi:10.1111/maq.12418.

Clarsen, G. (2017a). '"Australia – drive it like you stole it": Automobility as a medium of communication in settler colonial Australia'. *Mobilities*, 12(4), 520–533.

Clarsen, G. (2017b). 'Revisiting "Driving while black": Racialized automobilities in a settler colonial context'. *Mobility in History: Yearbook of the International Association for the History of Transport, Traffic & Mobility*, 8, 47–55.

Clarsen, G. (2002). 'Still moving: Bush mechanics in the Central Desert'. *Australian Humanities Review*, 25. Retrieved from www.australianhumanitiesreview.org/archive/Issue-March-2002/clarsen.html.

Cresswell, T. (2010). 'Towards a politics of mobility'. *Environment & Planning D: Society & Space*, 28(1), 17–31.

Davis, T. (2011). 'Darlene Johnson on making films in Arnhem Land' [Interview]. *Screening the Past*, 31. Retrieved from www.screeningthepast.com/2011/08/darlene-johnson-on-making-films-in-arnhem-land/.

Davis, T. (2006). 'Working together: Two cultures, one film, many canoes'. *Senses of Cinema*, 41. Retrieved from http://sensesofcinema.com/2006/feature-articles/ten-canoes/.

Davis, T. & Moreton, R. (2011). 'Working in communities, connecting with culture. Reflecting on U-matic to YouTube: A national symposium celebrating three decades of Australian Indigenous community filmmaking'. *Screening the Past*, 31. Retrieved from www.screeningthepast.com/2011/08/"working-in-communities-connecting-with-culture"-reflecting-on-u-matic-to-youtube-a-national-symposium-celebrating-three-decades-of-australian-indigenous-community-filmmaking-2.

De Heer, R. & Djigirr, P. (Dirs.). (2006). *Ten Canoes* [Motion picture]. Australia: Palace Films and Cinemas.

Dempsey, K. (2017). *In harm's way: A study of Northern Territory linked crash records*. Unpublished PhD Thesis, Charles Darwin University, Darwin, AU.

Hinkson, M. (2005). 'New media projects at Yuendumu: Towards a history and analysis of intercultural engagement'. In L. Taylor, G. Ward, G. Henderson, R. Davis & L. Wallis (Eds.), *The power of knowledge: The resonance of tradition* (pp. 157–168). Canberra, AU: Aboriginal Studies Press.

Konishi, S. (2011). 'Representing Aboriginal masculinity in Howard's Australia'. In R. Jackson & M. Balaji (Eds.), *Global masculinities and manhood* (pp. 161–185). Urbana, IL: University of Illinois Press.

Langton, M. (2003). 'Aboriginal art and film: The politics of representation'. In M. Grossman (Ed.), *Blacklines: Contemporary critical writing by Indigenous Australians* (pp. 109–124). Carlton, AU: Melbourne University Press.

Massey, D. (2005). *For space*. London: Sage.

Paul, M. & Bolognese, M. (2017). *Bush mechanics: From Yuendunu to the world*. Adelaide, AU: Wakefield Press.

Rose, D. (2017). 'Country and the gift'. In J. Adamson & M. Davis (Eds.), *Humanities for the environment: Integrating knowledge, forging new constellations of practice* (pp. 33–45). London: Routledge.

Rothwell, N. (2017). 'Colonial turbulence in the North'. *Arena Magazine*, 148, 11–17. Retrieved from https://search.informit.com.au/documentSummary;dn=950727169238332;res=IELHSS.

Rothwell, N. (2006, May 27). 'The new wave: How Ten Canoes is leading the revolution on the representation of Indigenous Australians on film'. *The Australian*.

Scott, R. & Heiss, A. (2015). *The Intervention: An anthology*. Sydney, AU: Concerned Australians.

Sheller, M. (2016). 'Uneven mobility futures: A Foucauldian approach'. *Mobilities*, 11 (1), 15–31.

Simpson, A. (2014). *Mohawk interruptus*. Durham, NC: Duke University Press.

Taylor, A. (2012). 'Information communication technologies and new Indigenous mobilities: Insights from remote Northern Territory communities'. *Journal of Rural & Community Development*, 7(1), 59–73.

12 Exploring the mobilities of forced displacement and state violence against homeless citizens in Bogotá, Colombia

Amy E. Ritterbusch

Introduction

On May 28th, 2016, homeless drug users and inhabitants of El Bronx, a drug consumption zone (*olla*) in the city center of Bogotá, Colombia, were violently displaced by the police as part of a state-sanctioned urban renewal initiative. CPAT and PARCES, two Colombian human rights NGOs supported by the Open Society Foundations, conducted an investigation of the intervention and authored the human rights shadow report *Destapando la Olla* (Tovar et al., 2017). The PARCES outreach team, which I led during the fieldwork for the report, conducted 20 semi-structured interviews and additional focus groups with Bronx exiles and employed participant observation in street spaces they occupied. The purpose of *Destapando la Olla* was to conserve the social memory of an urban community that seems doomed to erasure, as well as to generate public outcry and mobilize action against the perpetrators of this injustice against homeless citizens. The report was successful in focusing the attention of media and international human rights organizations on homeless people's lives and voices, their social and mobility rights as citizens and, specifically, the violation of their right to the city of Bogotá. It was also highly contested, and the validity of our methodologies was questioned in a counter-report issued by the Department of Economics at my university, the Universidad de los Andes (see Ritterbusch, 2018; Tobón & Zuleta, 2017); the latter was informally 'commissioned' by the local administration to delegitimize our report's documented claims of human rights violations and state violence against homeless citizens.

Although my colleagues and I endured efforts by other academics to undermine our findings, and exposed ourselves to state surveillance, I stand by the human rights principles underpinning our work. As other scholars have noted (Cahill, 2007; Pain, 2004), centering scholarly endeavors on social justice can be a troubling and uncomfortable path with devastating consequences for academics' personal and professional lives. Nevertheless, I believe that if we witness evidence of human rights violations, it is our duty as citizens and social justice activists to denounce them, even at personal cost. Accordingly, before closing its doors in December 2017 and despite efforts to dissuade our opposition to this instance of mobility injustice, PARCES led the case against the Colombian government in relation to these injustices; the case remains under consideration by the Inter-American Human Rights Courts (Ritterbusch, 2017).

The findings published in the PARCES report serve as a window on the world of injustice that Bogotá's homeless citizens have experienced for decades, and they created an uproar in Colombian social and mainstream media outlets. As the incident gained public attention, rumors and later evidence emerged in mainstream media outlets that homeless citizens had subsequently been herded onto trucks in the middle of the night and moved to other cities and small towns in Colombia. In this chapter I draw from (a) my experience as a PARCES member and first-hand witness of police violence, (b) testimonies cited in the *Destapando la Olla* report and (c) a range of mainstream news representations in order to examine both stages in this instance of involuntary movement: the expulsion of homeless citizens *from* El Bronx and their forced relocation *to* other towns and cities. By focusing on patterns of movement, media representations and lived experiences, the chapter provides an urban Latin American example of state domination through forced mobility (i.e., mobility injustice) that constitutes an infraction of homeless citizens' right to dwell and occupy public urban space in the name of security and progress in the city (i.e., social injustice).

In the following section I situate the chapter's focus in relation to recent work on mobility justice, especially its treatment of "domination in the political field of mobility" (Cook & Butz, 2016, p. 403). Drawing inspiration from Cresswell's influential work on the "politics of mobility", I then devote the paper's main section to describing "patterns of movement, representations of movement and ways of practicing movement" (Cresswell, 2010, p. 18) as they relate to the forced mobilities of homeless citizens in Colombia. Because my discussion relies largely on newspaper reports and homeless citizens' testimonies, and in keeping with Cresswell's (2010) insight that mobilities exist in the *entanglement* of movements, representations and practiced experiences, it is impossible fully to separate patterns and practices of movement from their representation in media and testimony. By using media excerpts to illustrate how state domination through forced mobility is represented in mainstream media outlets, my analysis aligns with mobility scholarship that highlights the importance of "popular culture in understanding the cultural constitution of mobility, and the way it is situated within political ideologies of particular geographical and historical conjunctures" (Shi & Collins, 2018, p. 127). The chapter concludes by reflecting on the measures that would be required – beyond activism in the representational realm – to achieve mobility justice for Bogotá's homeless citizens.

Social justice in the context of forced mobilities

Over the last decade, critical mobility scholars have pointed to the importance of analyzing the intersection of social justice and spatial mobility (Adey, 2006; Cook & Butz, 2016; Cresswell, 2010). For example, Cook and Butz

(2016, p. 400) interrogate "the entanglements of power, social exclusion and mobilities that articulate the intersection of mobility and justice. Who is free to move? Who is forced to move? Who is stuck in place?" They engage Sheller's (2012, pp. 199–200) claim that "ensuring mobility justice will entail moving beyond simply a politics of the *de jure* right to mobility, to instead ensuring that the *de facto* capability of mobility is protected and extended as a common basis for social justice". When considering the forced mobilities of homeless citizens in Bogotá, it is pertinent to interrogate homeless (im)mobilities as "both a political question (what rights to [im]mobility exist in a particular context, and how are they exercised and protected) and an ethical question (what capabilities of mobility are valued, defended and extended to all involved)" (Sheller, 2012, pp. 186–187). Using this politico-ethical framing, I explore the mobility injustices experienced by homeless citizens as a result of state actions that violently forced them to move throughout the city and to other cities, thereby negating their right to dwell.

Sheller (2012, p. 186) describes how "an approach highlighting mobility justice ... focuses our attention on who is able to exercise rights to [im]mobility, and who is not capable of mobility within particular situations". Cook and Butz push Sheller's thinking by more explicitly grounding mobility justice in social justice literatures. Drawing from Young's classic work on social justice, they describe how her theory of domination

> leads us to conceptualize mobility justice not only as the equitable distribution of motility throughout a social system, but also as the elimination of domination in the political field of mobility, or in other words as just institutional actions and decision-making processes about mobility issues that promote just mobility outcomes.
>
> (Cook & Butz, 2016, p. 403)

Cook and Butz engage Young's idea of domination to conceptualize unjust state actions and processes in relation to mobility:

> When a state makes decisions in a non-consultative manner behind closed doors, keeping citizens ignorant of tabled proposals and decision-making outcomes, it actively frustrates direct public participation in decision-making procedures. State authorities are able to discipline citizens by proscribing action that serves their own interests and by punishing behavior that threatens their decision-making authority. In this institutional context, citizens live within structures of domination.
>
> (Cook & Butz, 2016, p. 403)

The police raid on El Bronx is a clear example of this state domination in the field of mobility. The questions raised by mobility scholars about the connections between social (in)justice and spatial mobility are particularly relevant when considering this case of state-enforced mobility of homeless citizens within and between cities. Across multiple national contexts, scholars have documented and analyzed the forced mobilities of numerous communities, often characterized by conflict-driven movements between countries. Kleist's (2017) study of the narratives of involuntary return migration and cases of forced relocation to Ghana and Moret's (2017) analysis of mobility capital and the 'cross-border' mobility practices in the lives of former refugees and Somali migrants living in Europe are noteworthy examples. In concert with these studies, the case of forced movement of homeless citizens in Bogotá contextualizes how "there is as much un-freedom in mobility as there is in fixity" (Gill, Caletrío & Mason, 2011, p. 304).

Other work in the growing field of mobilities research reinforces this connection between mobilities and forced migration and calls for analyses of the mobilities of political refugees or victims of environmental displacement, cases in which mobilities are imposed by the state, a natural disaster or other actors or emergencies that force the movement of the most socially excluded citizens (Adey, 2016; Gill, Caletrío & Mason, 2011; Hannam, Sheller & Urry, 2006). Cresswell's (1999) work, for example, connects homelessness, embodiment and mobility in the context of female travelers in the United States, and centers on the gendered representations and moral panics surrounding their movement and presence in public space. While homelessness in the disciplines of geography and sociology is often analyzed in terms of representations of homeless spaces (e.g., Johnsen, May & Cloke, 2008) or state violence exerted against homeless people in these spaces (e.g., Stuart, 2015), this chapter (a) frames the forced movement of homeless citizens in Bogotá *from* their spaces of daily livelihood as an act of state domination and mobility injustice through which homeless citizens are denied their right to dwell in the city (cf. Sheller, 2016) and (b) explores the consequences of these acts of state domination through an examination of stigmatizing media representations of homeless movement that perpetuate violence against homeless citizens within both the spaces they are expelled from and the spaces to which they arrive in search of refuge.

Mobility injustice through state domination: the expulsion of homeless citizens from El Bronx and their forced relocation to other cities

The Bogotá police forcibly removed homeless citizens from El Bronx on May 28th, 2016, and in the days following herded them to a canal (Figure 12.1) located 18 blocks to the southwest. In order to control and contain homeless mobilities and drug use within this space, the police arrived at different times throughout the day and night. Sometimes two or three officers patrolled the area on foot; at other times they arrived on motorcycles in groups of 20 or more. During the weeks that homeless individuals were contained at the canal, they were violently moved within it every third night. Some nights they were pushed a few blocks east; on others, a few blocks west. During these forced displacements, the homeless experienced multiple forms of police

Figure 12.1 The canal on 6th Street, Bogotá
Source: Photograph by PARCES outreach team.

violence, including physical and verbal attacks, death threats and the use of
tear gas.

The canal is located on 6th Street in the center of Bogotá, and it is positioned below a congested Transmilenio station (Bogotá's rapid transit bus
system). For the weeks that homeless individuals were contained in the canal,
hundreds of passersby gazed down on this public health crisis of homeless
drug users, many of them visibly injured and without access to basic sanita-
tion and potable water. During observational activities, PARCES documented
reactions from the public, ranging from the indifferent gaze of commuters to
direct violent actions by residents from surrounding *barrios* who apparently
offered the homeless poisoned food. These initial patterns of movement
instigated by police – expulsions from El Bronx, relocation to the canal and

forced movement up and down it – are mechanisms of state domination that deny the rights of homeless citizens to dwell and occupy public spaces in the city. State domination through forced mobility has characterized drug users' experiences of the city for decades, and this echoes a pattern of social cleansing of street-connected communities, including homeless individuals, drug users and sex workers, across Latin American cities.

In their reporting on the operation, Bogotá's mainstream news media left no doubt that this program of forced movements was intended to effect the elimination of homeless citizens from Bogotá's public spaces. For example, one news piece described the forced displacement as follows:

> This Saturday early morning [around 3 a.m.], the authorities completed various interventions in the sector of El Bronx in Bogotá. ... "[T]he principal objective of the intervention in this sector is the protection, guarantee and re-establishment of the rights of minors", assured President Juan Manuel Santos. ... The mayor also announced that there are special urban renewal projects in this part of the center of Bogotá where in the future they may place some offices of the government entities and where they will place one of the metro stations. "All the tourists come to Bogotá and visit the city center, and it is important that it be clean and in order and not a sector where the visitors get attacked", stated Peñalosa [Bogotá's current mayor].
>
> (RCN News, 2016)[1]

Bogotá's mayor, Enrique Peñalosa, indirectly refers to inhabitants of El Bronx as those who attack tourists, and he employs the framing of heightened security as a principal justification for the intervention and subsequent acts of state-enforced movement of homeless citizens.

An article published in the newspaper *El Pais* shortly after the intervention presents a similar explanation by describing children's rights and the security state as the principal justification for the raid on El Bronx:

> 2,500 members of the public forces intervened in the zone early Saturday. ... There were two captures: alias TEO and alias "El Flaco" [both micro-traffickers]. Additionally, 24 children were rescued and attended to by 250 representatives of the Colombian Institute of Family Welfare – among the rescued there was a two-year-old boy. A building where the gang "Gancho Mosco" hid weapons, drugs and money was also discovered; two other gangs were also affected including "Gancho Manguera" and "Gancho Payaso"; and they rescued a man who was tied up by his feet and hands and with a chain around his neck. This intervention today in El Bronx is a totally coordinated intervention, an example of how collective work and coordinated work can produce good results. It is an intervention that has two objectives: one, to re-establish the rights,

over all else, of the children and of the citizens whose rights were being violated there; but also of course there was an objective to fight against crime because El Bronx became converted into a source of crime of all types. ... [The idea] was to take advantage of the intervention to combat drug trafficking and complete urban renewal in this zone, that is ironically located in front of where the Army does its recruitment exercises, and this will without a doubt make Bogotá a safer city.

(*El Pais*, 2016)

The main message is that a 'crackdown on crime' and the elimination of El Bronx will make Bogotá safer. Indeed, several criminal networks were engaged in illicit activities within, but not beyond, the boundaries of El Bronx. It was homeless citizens who left the boundaries of El Bronx to find food and/or money for drugs.

In the following excerpt, the daily newspaper *El Tiempo* praises the intervention as a crackdown on criminal organizations and the rescue of sexually exploited women and children:

The intervention, without precedent, was against the criminal organizations committing crimes in this sector. ... [T]he planning phase took four months and was led by the Sub-Secretary of Security, Daniel Mejia. ... [A]s a result of Saturday's intervention, the administration achieved important discoveries, including a man who had been kidnapped and chained in one of the buildings in the sector. ... Members of the Child and Adolescent Division of the Police and child and family welfare officials found nearly 200 women being sexually exploited, including 76 minors, 8 of whom were younger than 13 years.

(*El Tiempo*, 2016a)

The headline for this piece included a subtitle that focused on the rescue of kidnapped individuals and sexual slaves in El Bronx. Although these rights abuses and forms of violence did occur within El Bronx, the violation of homeless citizens' rights to linger and occupy public space are not mentioned, thus discursively privileging certain citizens' human and mobility rights over others. Without defending these criminal acts, the shadow report aimed to differentiate between organized crime occurring within El Bronx and the actions of homeless citizens living and doing drugs there. Media representations of these two groups are often conflated, which contributes to the criminalization of homeless citizens.

In the following passage from the weekly news magazine *Semana*, homeless citizens are represented as perpetrators of violence:

Since the administration of Enrique Peñalosa decided to assume control of El Bronx to put to an end to the most feared drug consumption zone in the city, the homeless have become an issue that doesn't seem to have a solution in Bogotá. Yes, if these people are visible victims of a dark business and of many social problems, they have also been protagonists of civil unrest and assaults in various zones of the city center. Their extreme vulnerability became evident this week when rising water dragged a group that was sleeping at the canal on 6th Street. Fortunately, firefighters were able to rescue some further downstream. Bogotá citizens increasingly feel terrified by their presence in many parts of the city. The question that many are asking is why the city and the police don't do something to take them to rehabilitation facilities or outlaw their presence in certain places? ... [P]art of this population are victims but also perpetrators, because in order to maintain their addiction they must commit crimes and violate the rights of other people. ... Can these people be taken to sites of medical emergency or, in the case of a diagnosis of mental illness, be treated or remitted without their consent?

(*Semana*, 2016)

The passage demonstrates how even the most prominent media outlets discursively dehumanize homelessness, which helps to justify acts of state domination and control over homeless bodies and movement. The excerpt even suggests devising a legal mechanism that permits such domination of decision-making processes pertaining to addiction and movement, despite international legislation outlawing forced rehabilitation and resettlement.

The news media also had much to say about the forced movement of homeless citizens *between* cities. Several articles claimed that the Bogotá administration moved homeless citizens – at the city's expense – to other cities as a form of expulsion, with negative implications for receiving municipalities. Testimonies from homeless individuals and state officials from the latter confirm this intercity movement and claim that it was instigated by the Bogotá government, although Bogotá administrators deny involvement.

Several news articles reported homeless individuals' testimony that they were deliberately sent to other cities by bus. According to Jhon Jairo Lemus, Pereira's Secretary of Social Development, homeless people arriving in Pereira stated that "they attended to us, they put us on the bus and they gave us food so that we would leave" (Caracol, 2016a). On its YouTube channel, Caracol News also published the testimony of a homeless individual who stated that "[Bogotá officials] took us to the bus station and they put us on buses". The news organization confirmed that a public official supported this homeless citizen's version of events (Caracol, 2016b). Human rights official Sandra Cárdenas, who had direct conversations with homeless citizens in Pereira, confirmed their arrival from Bogotá, stating that they

corroborated that they arrived in Pereira after the intervention of El Bronx, and that the Bogotá administration provided them with 200,000 pesos [US$70] to return to their cities of origin. The situation here is that those who arrived are not necessarily from Pereira ... they heard about the good weather and the good nature of the people. ... We haven't been able to count them because they are a mobile population and they don't have identification cards.

(Caracol, 2016b)

Caracol Radio and *El Espectador* also cited officials from Pereira and Dosquebradas who voiced concerns about their arrival and whereabouts. According to *El Espectador* (2016),

At least 300 homeless individuals from Bogotá arrived to Pereira [a small city northwest of Bogotá in the *Eje Cafetero* (coffee zone)] after the intervention in El Bronx. This accusation was presented by Jhon Jairo Lemus, Secretary of Development of the Administration of Pereira, and Sandra Cárdenas, Human Rights district official. Both officials assured, in their conversation with this news source, that many homeless individuals said that the Secretary of Social Integration of Bogotá paid for their trip. Various individuals claimed that they had received additional money for the displacement. ... Cardenas stated that the arrival of homeless individuals to the city started a week after the intervention in El Bronx. ... Since then, both officials have perceived an increase in levels of insecurity.

Pereira city officials evidently denounced the Bogotá administration's act of domination through forced mobility, and they described the implications of this displacement for their own city. Other political leaders provided further details about the impact of the displacement and resettlement for local citizens and surrounding communities:

The community of Pereira continues to worry about the massive presence of homeless individuals, especially in the center of the city. ... [M]any of the citizens discuss how, because of this, robbery has increased and many residential zones are no longer safe. The citizens that live between Liberty Park and Flag Park and between Avenues 6 and 10 confirm being victims of robbery and intimidation by these individuals, who have taken control of different residential zones to sleep, go to the bathroom and beg for food during the day.

(Caracol Radio, 2016)

Secretary Lemus focused on the increase in homelessness in Pereira:

Jhon Jairo Lemus assured us that the current census on homelessness indicates an increase, and that the count is over 1,300 homeless individuals. He also indicated that every day there are more homeless individuals. ... "At the beginning of this administration the count was 927 homeless individuals, and currently we have between 1,300 and 1,400 homeless individuals. They move around between *Dosquebradas, Manizales* and *Armenia* [three other small cities and towns in the *Eje Cafetero*]. We are in the process of characterizing and identifying their origin", said Lemus, who confirmed that at least 2,000 homeless individuals left Bogotá and came to the *Eje Cafetero*.

(Caracol Radio, 2016)

One month later, Lemus declared that Pereira was in a state of emergency because of the increase in homeless individuals:

If you live in the Risaralda capital [Pereira], you will have noticed that the number of homeless individuals has increased. It is not a perception, it is reality. ... [I]t is clear that many [from El Bronx] are in the city "because they have completely different characteristics than the homeless individuals from here". "It is an emergency". ... "We declare that we are in a perfect emergency because of the problem of homeless individuals, and we request support from the national government", stated the official. ... Lemus requested that Pereirians not give out money for any reason to these people. He warned that some of the recently arrived homeless individuals had very aggressive behaviors and also intimidate others in order to be given money. They commit robbery of individuals and vehicles.

(*El Tiempo*, 2016b)

Lemus also remarked publicly that

At least 2,000 homeless individuals traveled from Bogotá, pushed by entities such as the Secretary of Social Integration that supposedly covered their transportation expenses. They put them on a bus to Pereira, and they gave them a food subsidy for the trip.

(Kienyke, 2016)

Press coverage describing the movement of homeless citizens between cities was stigmatizing and dehumanizing:

Some left the city for municipalities on the outskirts of Bogotá. In the first days of June, the Mayors of Cajicá, Chía, Zipaquirá, Madrid and Mosquera announced that homeless individuals different from the local community were arriving at their municipalities. ... "They said that they arrived here in trucks, in unofficial cars and that they were left on the highway". ... [T]he Secretary of Government of Cajicá assured [local residents] that they returned the homeless to Bogotá in police patrol cars. Then they started appearing in Ibagué, Pereira, Dosquebradas, and Villavicencio: cities that are located in three different departments. The municipalities suspect that the homeless individuals come from El Bronx.

(Las 2 Orillas, 2016)

The headline of this piece reads: "The remedy has been worse than the illness" (Las 2 Orillas, 2016). These media excerpts represent the repressive actions of the Bogotá government as merely displacing this social dilemma to another urban context. Additionally, as demonstrated in statements from Pereira officials, the local governments of receiving cities also framed the issue of homelessness in security terms rather than focusing on the right of these individuals to dwell and occupy public spaces in cities throughout the country.

But how do these media representations of homeless people's forced displacement and relocation compare to the victims' own telling of their experiences? As might be anticipated, while municipal press releases and news media reports framed the forced expulsion and relocation either as a necessary and rationally executed way to clean up Bogotá's public spaces or as mere displacement of the 'problem' of the homeless from one municipality to another, homeless citizens' accounts of the intervention focus on its violence and inhumanity.

In official statements to the press, Bogotá officials deny inflicting violence on the homeless during their eviction from El Bronx and moving them to Pereira and other cities. However, homeless citizens' testimonies confirm acts of violent state domination. One homeless El Bronx exile described her experience of forced movement into the canal on 6th Street in Bogotá as follows:

> The police officer came after me on a motorcycle. That day it was approximately 3 or 4 o'clock in the morning, there wasn't a soul in sight ... When he caught up with me, he said "come here, come closer". And as soon as I got close he grabbed my arm and threw me into the wall, and if I hadn't been blocking my face, he would have smashed it in. And as soon as I raise my hand he twists my other hand, almost breaking my hand. ... For two days, I couldn't stand up and was vomiting blood. I am pregnant, I am pregnant, and for two days I couldn't stand, I threw up blood. And had chills, a fever and feeling pain in my body. ... Why? Because of the beating that cop gave to me that day, and besides that he came after me, on top of me with his motorcycle and made me run until 6th Street [to the canal location], from Cinco Huecos [a neighborhood and drug zone in the center of the city] until 6th [Street]. Running with the motorcycle practically on top of me.
>
> (Tovar et al., 2017, p. 51)

Another long-term inhabitant of the canal had this to say:

> This early morning they arrived, displaced us with a violent system, very violent. There were more than 80 police around the canal. They threw tear gas at my peers. They threw more than 20 shots of tear gas [into the canal] and they didn't let them leave the canal even though the water was

rising because it had been raining ... it was said that they opened up the flood gates of the canal ... and then 20 of our peers passed by floating in the water. And I had to provide a rope to pull a girl out who was crying for help, that she was drowning. That night was unbelievable. We didn't know that the canal had floodgates. We learned from the news channels when they came to ask us if we could confirm whether they had opened the floodgates. This is so strange, because I have been living in the canal for six years. I have never seen the water rise. ... The police would not let them leave the canal.

(Tovar et al., 2017, pp. 53–54)

Although news outlets were content to draw from victims' testimony to validate claims that Bogotá's government was behind the relocation of homeless citizens to other cities, they downplayed or ignored testimony relating to police brutality or violation of rights. From the perspective of social justice and mobility justice, which version of events should we believe? Do we prioritize media representations and politicians' voices or victims' accounts of forced movement? Principles of epistemic justice require that the latter be given due priority in order to avoid wronging members of Bogotá's homeless population "in their capacity as a giver of knowledge" and as "a subject of social understanding" (Fricker, 2007, p. 7; see also Butz & Cook, this volume). In the context of state violence and domination of particular citizens' (im)mobilities, privileging victims' experiences is especially important, as is attending to media silences and political denials. Further research is needed to enhance our understanding of such interventions and their justice-related effects on the ((im)mobile) lives and bodies of the homeless throughout Colombia.

Conclusion: just mobility outcomes for the homeless?

My objective has been to provide a case study of state domination through forced mobility by contextualizing the forms of violence and injustice experienced by homeless citizens during the 2016 intervention in Bogotá. They experienced social and mobility injustice as their right to dwell and occupy public space was infringed as an outcome of state-driven mechanisms of social control over their bodies and presence in Colombian cities. In this context of domination and forced movement, what would a "just mobility outcome" (Cook & Butz, 2016, p. 403) look like? Mobility justice would involve Colombian state institutional actions, mechanisms and decision-making processes that engage homeless citizens about the dynamics of their lives and guarantee their right to (im)mobility and access to the city without resorting to violent, repressive urban politics that move them 'out of sight' under the auspices of state security and citizen safety.

Through the interventions of PARCES, particularly our on-the-ground work as civil watchdogs in the 6th Street canal and the visibility we brought to this and related injustices through our local and international-scale human rights mobilizations, we were able to denounce the violence exerted against homeless citizens by Colombian government actors. However, the resulting situation for the homeless in Bogotá is still far from a just mobility outcome. Our strategies fell short in terms of preventing future acts of state domination through forced mobility. In addition to researching mobility injustice and mobilizing resultant data as tools in an international human rights case, we need preventative justice measures that safeguard homeless citizens' social and mobility rights over time. Those measures include changing institutional structures and street-level interactions between state actors and citizens in terms of restorative justice, processes through which historical damages are redressed. From my perspective as a witness to mobility injustice and mobility violence, justice for homeless citizens also should be realized through a formal legal process against the Colombian state, in addition to guaranteeing the basic rights of the homeless, including the right to dwell and to occupy public spaces in the city.

Note

1 All translations are mine.

References

Adey, P. (2016). 'Emergency mobilities'. *Mobilities*, 11(1), 32–48.
Adey, P. (2006). 'If mobility is everything then it is nothing: Towards a relational politics of (im)mobilities'. *Mobilities*, 1(1), 75–94.
Cahill, C. (2007). 'Repositioning ethical commitments: Participatory action research as a relational praxis of social change'. *ACME: An International E-Journal for Critical Geographies*, 6(3), 360–373.
Caracol. (2016a, September 2). 'Habitantes de la calle denuncian que los mandaron a Pereira desde Bogotá'. *Caracol News*. Retrieved from https://noticias.caracoltv.com/colombia/habitantes-de-la-calle-denuncian-que-los-mandaron-pereira-desde-bogota.
Caracol. (2016b). 'Habitantes de la calle denuncian que los mandaron a Pereira desde Bogotá' [Video file]. *Caracol News*. Retrieved from www.youtube.com/watch?v=bhZst2iPMBw.
CaracolRadio. (2016, September 1). 'Ya va en 1300 la cifra de habitantes de calle en Pereira' [Radio transmission]. *Caracol Radio*. Retrieved from http://caracol.com.co/emisora/2016/09/01/pereira/1472730623_426331.html.
Cook, N. & Butz, D. (2016). 'Mobility justice in the context of disaster'. *Mobilities*, 11 (3), 400–419.
Cresswell, T. (2010). 'Towards a politics of mobility'. *Environment & Planning D: Society & Space*, 28(1), 17–31.
Cresswell, T. (1999). 'Embodiment, power and the politics of mobility: The case of female tramps and hobos'. *Transactions of the Institute of British Geographers*, 24 (2), 175–192.
ElEspectador. (2016, September 1). '¿Éxodo de los habitantes de calle a Pereira?'. *El Espectador*. Retrieved from www.elespectador.com/noticias/bogota/exodo-de-los-habitantes-de-calle-articulo-652490-0.

ElPaís. (2016, May 28). 'Tras gigantesca intervención, Gobierno dice que "El Bronx" no volverá a ser el mismo'. *El Pais.* Retrieved from www.elpais.com.co/judicial/tra s-gigantesca-intervencion-gobierno-dice-que-el-bronx-no-volvera-a-ser-el-mismo. html.

ElTiempo. (2016a, May 28). 'Alcaldía encontró secuestrados y esclavas sexuales en El Bronx: Gigantesco operativo sorpresa con 2.500 hombres intervino el sector ubicado en el centro de Bogotá'. *El Tiempo.* Retrieved from www.eltiempo.com/a rchivo/documento/CMS-16605518.

ElTiempo. (2016b, August 6) 'Pereira en emergencia por aumento de habitantes de calle del "Bronx": Tras los operativos realizados en Bogotá, esta población está llegando a la ciudad'. *El Tiempo.* Retrieved from www.eltiempo.com/colombia/otra s-ciudades/pereira-en-emergencia-por-aumento-de-habitantes-de-calle-37152.

Fricker, M. (2007). *Epistemic injustice: Power and the ethics of knowing.* New York: Oxford University Press.

Gill, N., Caletrío, J. & Mason, V. (2011). 'Introduction: Mobilities and forced migration'. *Mobilities,* 6(3), 301–316.

Hannam, K., Sheller, M. & Urry, J. (2006). 'Editorial: Mobilities, immobilities and moorings'. *Mobilities,* 1(1), 1–22.

Johnsen, S., May, J. & Cloke, P. (2008). 'Imag(in)ing "homeless places": Using autophotography to (re)examine the geographies of homelessness'. *Area,* 40(2), 194–207.

Kienyke. (2016, September 2). '¿Del Bronx al Eje Cafetero? Denuncian traslado de habitantes de calle'. *Kienyke.* Retrieved from www.kienyke.com/noticias/habitantes-de-calle-bronx-eje-cafetero.

Kleist, N. (2017). 'Trajectories of involuntary return migration to Ghana: Forced relocation processes and post-return life'. *Geoforum.* doi:10.1016/j. geoforum.2017.12.005.

Las2Orillas. (2016, August 18). '¿Qué se hicieron los habitantes del Bronx?' *Las 2 Orillas.* Retrieved from www.las2orillas.co/que-se-hicieron-los-habitantes-del-bronx/.

Moret, J. (2017). 'Mobility capital: Somali migrants' trajectories of (im)mobilities and the negotiation of social inequalities across borders'. *Geoforum.* https://doi.org/10. 1016/j.geoforum.2017.12.002.

Pain, R. (2004). 'Social geography: Participatory research'. *Progress in Human Geography,* 28(5), 652–663.

RCNNews. (2016, May 28). 'Fuerte intervención en El Bronx deja como resultado tres bandas desarticuladas'. *RCN News.* Retrieved from www.noticiasrcn.com/naciona l-bogota/fuerte-intervencion-el-bronx-deja-resultado-tres-bandas-desarticuladas.

Ritterbusch, A. (2018). 'The quantitative tactics used to delegitimize a right to the city: Social justice movement in Bogotá, Colombia'. *Alternate Routes: A Journal of Critical Social Research,* 29, 254–267.

Ritterbusch, A. (2017, September 28). 'Who are the real targets of Bogotá's crackdown on crime?' *The Conversation.* Retrieved from http://theconversation.com/who-a re-the-real-targets-of-bogotas-crackdown-on-crime-83949.

Semana. (2016, August 21). 'Habitantes de calle: tremendo problema'. *Semana.* Retrieved from www.semana.com/nacion/articulo/habitantes-de-calle-tremendo-p roblema-en-bogota/489734.

Sheller, M. (2016). 'Uneven mobility futures: A Foucauldian approach'. *Mobilities,* 11 (1), 15–31.

Sheller, M. (2012). 'The islanding effect: Post-disaster mobility systems and humanitarian logistics in Haiti'. *Cultural Geographies,* 20(2), 185–204.

Shi, Y. & Collins, F. (2018). 'Producing mobility: Visual narratives of the rural migrant worker in Chinese television'. *Mobilities*, 13(1), 126–141.

Stuart, F. (2015). 'On the streets, under arrest: Policing homelessness in the 21st century'. *Sociology Compass*, 9(11), 940–950.

Tobón, S. & Zuleta, H. (2017). 'Comments on the document – Uncovering the "*Olla*": Shadow report on the intervention in El Bronx', prepared by CPAT and PARCES. Bogotá: CESED, Universidad de los Andes.

Tovar, M., Trejos, C., Giraldo, Y., Delgado, G., Lanz, A., Lanz, S., León, S., Lloreda, A., Pardo, L., Morales, A. & Salamanca, J. (2017). *Uncovering the "Olla": Shadow report on the intervention in El Bronx*. Bogotá: Impresol Ediciones.

13 LGBTQ communities, public space and urban movement

Towards mobility justice in the contemporary city

Catherine J. Nash, Heather Maguire and Andrew Gorman-Murray

Introduction

This chapter examines the relationship between 'just' public urban space and mobility justice for LGBTQ communities in Toronto, Canada, and Sydney, Australia.[1] The mobilities considered are movements across and between neighborhoods in these cities by LGBTQ people. We argue that mobility justice is contingent upon creating more accessible urban space for all citizens, accounting for the social diversity of the city, including its LGBTQ communities. The relationship between urban spaces and mobilities is thus iterative: mobility justice is predicated on having access to public spaces in and through which to move freely, and that movement also produces and shapes just urban spaces. This relationship is processual. As we have shown in earlier work (Gorman-Murray & Nash, 2014; Nash & Gorman-Murray, 2015a, 2015b), recent social, political and legal gains have enabled some LGBTQ people to move more freely and visibly across wider spectra of Toronto and Sydney than in the past, when they were 'closeted' in private spaces, red-light districts and gay villages. These gains include equality in marriage, family formation and succession law and protection through anti-discrimination legislation. We have argued that this new social, political and legal landscape has engendered 'new' queer mobilities and contributed to a reshaping of LGBTQ inner-city neighborhoods.

We extend this analysis by thinking about these changes in terms of 'justice' – just public urban spaces and mobility justice. We draw on Low and Iveson's (2016, p. 16) evaluative framework for "the justice of public spaces" to analyze the relationship between just public urban spaces and mobilities. This framework draws out five lenses of social justice in urban life and its diversity: public space and distributive justice; public space and recognition; public space, encounter and interactional justice; public space and care and repair; and public space and procedural justice. We begin with a discussion of scholarship on sexuality, gender and urban spaces with a focus on the emergence of so-called gay villages as political, social and economic hubs of

LGBTQ urban life in many mid- to large-sized cities in the Global North. We then turn our attention to how the 'new mobilities' approach can be utilized to understand how changing LGBTQ mobilities are reshaping neighborhoods and access to public space in Toronto and Sydney. Next, we employ Low and Iveson's (2016) five lenses of justice to evaluate how these changes have materialized in more just urban spaces and mobilities.

LGTBQ communities and public space

Geographical scholarship on sexuality and public urban space dates back to the early 1980s and highlights the role of LGBTQ communities in urban place-making. In the 1980s and 1990s, scholarship detailed the development of urban LGBTQ residential and commercial neighborhoods – admittedly dominated by gay men – and thus constituting localized LGBTQ communities. These 'gay villages' have come to represent, both materially and symbolically, LGBTQ political, social and economic strength in inner cities (Castells, 1983; Knopp, 1990; Nash, 2005, 2006). Gay villages also serve as bulwarks against mainstream opprobrium and as bases to contest often-violent exclusion from most public spaces. Pride celebrations, for example, highlight deliberate attempts to 'queer' public space to contest heterosexual norms (McCartan, 2017). With the rise of neoliberal processes that reshaped post-industrial, entrepreneurial cities, gay villages were enlisted to buttress cities' claims about their tolerance, diversity and acceptance of difference (Bell & Binnie, 2000; Binnie, 2004).

Scholarship also details the role of urban space in the constitution and disciplining of normative sexual and gender identities, and the resulting exclusion of LGBTQ people in mainstream public spaces (e.g., Nash, 2005; Valentine, 1995). Normative social identities and subjectivities, disciplined through the dominant norms and expectations embedded in public spaces, ensured that visible lesbian and gay identities were rendered 'out of place', leading to the marginalization of visible queerness through self-exclusion or verbal or physical violence (Browne, 2007; Nash, 2005, 2006). Together, this scholarship illustrates how an LGBTQ presence in public space is closely regulated, but also reinforces the notion that LGBTQ people are more properly 'in place' in gay villages.

Despite the central role of gay villages in LGBTQ life, recent scholarship argues that in some cities, gay villages are in decline (Collins, 2004; Nash, 2013b; Ruting, 2008). Gay villages are frequented by a broader array of consumers and residents, with many commercial spaces becoming more mixed or exclusively heterosexual (Collins & Drinkwater, 2017). Further, given social, political and legal gains, LGBTQ people are increasingly comfortable (and accepted) in other inner-city venues as well as in some suburban and rural locations (Gorman-Murray, Waitt & Gibson, 2012). Newer generations of so-called post-gays increasingly reject sexual orientation as a predominant marker of identity and, through social media, are able to find other like-minded individuals without needing to frequent expressly LGBTQ spaces (e.g., Nash, 2013b; Sullivan, 2005).

LGBTQ mobilities

In both Toronto and Sydney, empirical evidence supports claims that local gay villages are in decline and that newer, alternative LGBTQ places and neighborhoods are evolving (Gorman-Murray & Nash, 2014, 2017). As a result of new social, political and legal status, some LGBTQ people enjoy greater access, visibility and mobility across inner-city neighborhoods than previously experienced (Nash & Gorman-Murray, 2014, 2015b; Gorman-Murray & Nash, 2014, 2017). These new mobilities constitute new relational geographies (and new subjectivities for some) through "the social content of movements of people and objects from place to place at various scales and the immobilities and 'moorings' that underpin and challenge these dynamics" (McCann & Ward, 2011, p. 176). Emerging LGBTQ places represent new moorings (or stabilizations) of particular objects, knowledges, people, goods and things – highlighting the unstable and sometimes temporary formations of new relational geographies (Gorman-Murray & Nash, 2014, p. 626).

The concept of relational geographies has allowed us to develop a more nuanced and complex assessment of the evolution, instability and formulation of new LGBTQ urban spaces (Gorman-Murray & Nash, 2014; Nash & Gorman-Murray, 2014). Rather than simply suggesting that gay villages are in decline and other locations on the rise, we contend that emerging relational geographies position gay villages within a broader affiliation of inner-city networks that constitute a new LGBTQ inner-city landscape, comprising LGBTQ places and neighborhoods along with the transport infrastructure and geographical imaginaries connecting them. These new urban spaces and mobilities are also unfolding differently in Toronto and Sydney: in Toronto, there is perhaps a more diffuse network; in Sydney, a greater weathering of the gay village in Sydney and the fluorescence of a visible LGBTQ neighborhood centered on Newtown (Gorman-Murray & Nash, 2017; Nash & Gorman-Murray, 2015a, 2015b).

In making these arguments, we recognize that urban spaces and mobilities are socially produced and, therefore, embedded in and operating through social power relations. Recognizing power relations points to the "politics of mobility" defined by (Cresswell, 2010, p. 21):

By politics, I mean social relations that involve the production and distribution of power. By a politics of mobility, I mean the ways in which mobilities are both productive of such social relations and produced by them. Social relations are of course complicated and diverse. They include relations between classes, genders, ethnicities, nationalities, and religious groups as well as a host of other forms of group identity.

Cresswell signals the need to pay attention to who (or what) gets to move, who does not and how those movements are governed and surveilled in the production of new relational geographies. Unequal social relations constituted through understandings of gender, race, sexuality and age, amongst others, frame the possibilities and experiences of mobilities within certain historical and geographical contexts (Adey, 2006; Cresswell, 2010). 'Motility' (Kaufmann, 2002) – the 'potential' for mobility – requires researchers to consider who can be mobile and in what circumstances, such that some people have more 'mobility capital' than others (Dufty-Jones, 2012; Jensen, 2009). Different realities and capacities for urban mobilities raise questions about the 'justness' of access to, and experiences in, new LGBTQ urban spaces emerging in Toronto and Sydney (Cook & Butz, 2016; Sheller, 2016). The concept of 'mobility justice' helps us better understand changing LGBTQ experiences of urban spaces and mobilities in contemporary Toronto and Sydney.

LGBTQ communities and mobility justice in the city

Mobility scholars are concerned with how uneven mobilities and motilities limit some people's capacity to fully participate in political, economic or social life, raising questions about mobility justice (Adey, 2006; Cook & Butz, 2016; Jensen, 2011; Kaufmann, 2002; Uteng & Cresswell, 2016). Sheller (2014) argues that the concept of mobility justice requires researchers to engage with the "different relations around mobility, ... the power differentials that come into play in any form of mobility and the different affordances that different people are able to make use of, or appropriate, in becoming mobile or not". This literature considers mobility justice in terms of gender (e.g., Cook & Butz, 2018; Uteng & Cresswell, 2016), sexuality (e.g., Montegary & White, 2015), race (e.g., Nicholson and Sheller, 2016) and disability (Goggin, 2016). Research has also explored accessibility to transportation networks (Culver, 2017), health and safety (Freund & Martin, 2007), pollution and emissions (Mullen & Marsden, 2016), environmental justice (Mullen & Marsden, 2016; Sheller, 2011) and an overall move towards sustainable development (Sheller, 2011). Sexuality, in the context of mobility justice, remains largely unexplored.

Geographical scholarship argues that LGBTQ people have been severely constrained in urban public spaces and that considerable LGBTQ activism has contested these exclusions. As a general proposition, Low and Iveson (2016, p. 10) point out that "conflicts over access to public spaces [are] one of the defining features of contemporary urban politics". The term 'public space' includes a range of locations that are, ideally, open and accessible for the free movement of all people through fair and democratic governance (Low & Smith, 2006; Mitchell, 2003; Staeheli & Mitchell, 2007). Scholars argue that public space, understood in this way, is pivotal for the working of democracy, allowing social groups to present their ideas and engage in public debate (Mackintosh, 2017). Exclusion from public space can constitute exclusion from full and fair participation in social and political life. Public space is also highly valued for fostering engagement with unfamiliar 'others' in ways that promote social cohesion and conviviality (Fincher & Iveson, 2008; Gorman-Murray & Nash, 2017).

The constitution of new LGBTQ places and neighborhoods in Toronto and Sydney, predicated on greater legal and political acceptance, suggests that LGBTQ people are experiencing enhanced social and physical mobilities, reflecting enhanced mobility justice. To assess if these changes in LGBTQ urban spaces and mobilities do constitute some form of increased mobility justice, we use Low and Iveson's (2016, p. 16) five lenses for "evaluations of the justice in public spaces". We explore each of these lenses in turn, drawing on our work in Toronto and Sydney for examples.

Public space and distributive justice

Low and Iveson (2016, p. 16) understand distributive justice as "how the wealth, rewards, benefits and burdens of urban life should be distributed to achieve a just city". In considering public space and distributive justice, they argue for evaluating the geographical distribution of public spaces across the city, as well as the question of affordable access to those public spaces. Although there might be 'unequal' distribution of public spaces, to find such a distribution 'unjust', the unequal outcome must be the "product of processes that systematically produce and maintain inequalities" that favor some over others (Low & Iveson, 2016, p. 17).

For LGBTQ people, the concern is visibility and safety as a 'queer' person in public space, as well as the geographical distribution of LGBTQ places. Some LGBTQ people (particularly gay and bisexual men) have a long history of carving out queer public spaces for socializing and cruising, such as Hanlon Point in Toronto or Hyde Park in Sydney (Nash, 2006; Wotherspoon, 2016). In the 1970s, such locations were open secrets, policed sporadically and unpredictably by local authorities. As gay villages were consolidated in both cities, LGBTQ people developed a territorial base for socializing and protest. Nevertheless, in gaining concessions around public visibility in certain neighborhoods, LGBTQ people experienced feelings of exclusion from other public locations (e.g., commercial establishments) for being 'out of place' (Bain, Podmore & Rosenberg, 2017). These constraints on LGBTQ mobilities arose from "processes that systematically produce[d] inequalities" in public spaces in both Toronto and Sydney (Low & Iveson, 2016, p. 17).

Arguments about constraints in public spaces need to be further parsed to emphasize uneven mobilities within LGBTQ communities. From the 1960s onward, emergent gay villages were largely frequented by white, middle-class gay men with the financial means and mobilities to make use of these inner-city spaces (Nash, 2006; Wotherspoon, 2016). Lesbians and queer women have never been fully accepted in these 'gay' locations, being subject to discrimination and sexism, at the same time that women's presence in public spaces was regarded as problematic (Jennings, 2015; Nash, 2006).

LGBTQ mobility justice needs to be attentive to the gendered, racialized and classed exclusions from LGBTQ public spaces in assessing a reduction in the "processes that systematically produce inequalities" (Low & Iveson, 2016, p. 17). New LGBTQ spaces in Toronto, for instance, are often found in particular restaurants, bars and dance clubs that only some can afford. Many of these locations are temporary; they might hold a 'queer night' once a week, making them less accessible to LGBTQ people on other days. They might also be less friendly or supportive of people of color, lesbians, trans individuals or older LGBTQ people, and the neighborhood itself might not be particularly safe, making it riskier to take the bus or walk to venues (Nash, 2013a, 2013b). Nonetheless, LGBTQ people are accepted in a broader array of downtown locations, with visible LGBTQ populations in neighborhoods such as Parkdale in Toronto and Newtown in Sydney. Simultaneously, a new LGBTQ generation increasingly regards gay villages as outliving their usefulness and belonging to an 'older generation' (Reynolds, 2009). Yet this expanding LGBTQ inner-city constituency is arguably fueling gentrification, raising property values and displacing working-class populations (Nash, 2013b). While, as a group, LGBTQ people might have wider geographical access to the city, in many ways, with a reduction in unjust public spaces, these contemporary mobilities are for a favored few.

Public space and recognition

In their second lens, Low and Iveson (2016, p. 18) argue that assessing justice in public spaces by focusing on 'recognition' addresses "the systematic devaluing and stigmatization of some urban identities and ways of life in cities". Being visible as LGBTQ in public spaces is a matter of personal risk as LGBTQ people's "very identities and ways of being ... are unfairly denigrated or stigmatized" (Low & Iveson, 2016, p. 18). Low and Iveson (2016) argue that equality of recognition means working against "cultural patterns that systematically deprecate some categories of people and the qualities associated with them" (Fraser, 1996, p. 31). For LGBTQ people, displays of public affection and ways of dressing or speaking (especially if gender nonnormative) often result in exclusion from public space through intimidation and violence (Noack-Lundberg, 2012). This exclusion is sometimes grounded in claims about the need to protect children, by naming LGBTQ people as sexual predators or by suggesting their very presence might 'confuse' children. The police, bylaw officers and fire and liquor-licensing inspectors have acted as the arm of the state in LGBTQ exclusion from public space.

Greater mobilities have opened up public spaces and semi-commercial locations for LGBTQ participation. However, the same caveats apply with respect to who benefits from greater access and movement. Attention must be paid to the gendered, racialized and classed aspects of recognition. Being a 'woman' in some locations is problematic, depending on the time of day or night, or the nature of the location. Intersectional identities, recognized through or within distinctive racialized, cis-gendered, queer or trans embodiment, also need consideration in determining mobility justice. In other words, homonormative (and normatively gendered) gay men and women have greater access to public spaces even if they are recognized as 'queer' (Leonard et al., 2012; Noack-Lundberg, 2012).

Public space, encounter and interactional justice

Low and Iveson's (2016) third lens attends to moments of public encounter and interactional justice. Through participation in public spaces, people assess the justness or fairness of their encounters "based on the quality of interactional treatment they receive" (Low & Iveson, 2016, p. 18). 'Quality' involves an assessment of the variety and respectfulness of those encounters. The absence of abuse is significant and should foster increased cooperation and conviviality – a key purpose of public spaces in democratic society (Low & Iveson, 2016; see also Mackintosh, 2017). The justness of interactions works with a politics of recognition in considering the fairness and equitable treatment of people in specific public spaces. This third frame of justice highlights how the provision of some public spaces fosters more 'convivial' atmospheres allowing inhabitants to 'bump into' each other, converse and engage in ways that spark mutual understanding (Fincher & Iveson, 2008; Low & Iveson, 2016).

Low and Iveson's framing of the just city as one of positive encounter can be linked to scholarship on the use of public space to promote social cohesion and conviviality. Social cohesion in this context is "the absence of latent social conflict and the presence of strong social bonds" (Witten, McCreanor & Kearns, 2003, p. 323). Arguably, a more socially cohesive neighborhood supports myriad moments of encounter with difference, thereby improving people's comfort with such interactions such that the experience of difference is positive, which in turn renders these spaces more just. In Toronto, those working to promote the gay village argue that LGBTQ people are welcome throughout Toronto, while making claims about the village as an expressly LGBTQ space that is welcoming to all (Nash and Gorman-Murray, forthcoming). This reflects activities by LGBTQ communities (including businesses and political leaders) to create public spaces of sociability and cohesion to support economic development in the village, in part, but also to improve the politics of encounter and recognition (Gorman-Murray & Nash, 2017).

Safety from homophobic encounters is fundamental to just public spaces and mobilities. As Knopp (2007, p. 23) argues, historically "the visibility that placement brings" can make LGBTQ populations vulnerable to violence. Establishing new LGBTQ venues (coffee shops, bars, restaurants) is fraught given that such transformations can be highly contested. The production and maintenance of LGBTQ spaces, even for gay villages, renders them potential targets for violence and abuse (Gorman-Murray & Waitt, 2009). While cosmopolitan cities are often considered much safer than suburban or smaller cities, the risk of violence remains. In September–November 2017, Sydney's alternative queer neighborhood, Newtown, and other nearby inner-city neighborhoods, such as

Redfern, experienced homophobic attacks during and after the Australian Marriage Equality Postal Survey (Tilley & Hoad, 2017).

So while greater access and mobility for LGBTQ people can enable more just public spaces through encounter and interactional justice, the picture is more complicated. Legal and social changes mean that more positive public encounters are happening, leading to a much-improved sense of community, as difference becomes 'normalized' through quotidian interactions – but only for some queer people in certain locations. Moreover, while more inner-city locations are available to LGBTQ people, their movements remain constrained by the nature of these locations and particular gendered, classed and racialized subjectivities.

Public space and care and repair

Low and Iveson (2016), through their fourth lens, argue that focusing on care in, and repair of, urban public space requires looking beyond issues of recognition and interaction to examine how people attend to the needs of each other and of their environment. This requires us to think about the often-unacknowledged aspects of city life, from unpaid care for the sick and elderly to the maintenance and repair of urban spaces through cleanup initiatives that attest to the cooperative way spaces are maintained (Low & Iveson, 2016). Here we take a somewhat tangential approach to Low and Iveson's (2016) notion of care and repair by examining how the creation and preservation of the gay village reflects an ethic of care within the LGBTQ community.

When we consider the historical geographies of gay villages through the lens of landscapes of care and repair, gay villages are locations that establish and promote care for a specific community through cooperation and self-help. From the 1970s onwards, LGBTQ activists created and preserved spaces in the service of LGBTQ communities – spaces that underpin, or moor, LGBTQ mobilities. In Toronto, the 519 Church Street Community Centre, Glad Day Books, Buddies in Bad Times Theatre and the AIDS Committee of Toronto were all located in and around the gay village and created public and semi-public spaces serving the LGBTQ community (Nash, 2006; Warner, 2002).

In the last decade or so, concerns about the demise of the gay village have led to several initiatives to refurbish and/or maintain these spaces. After Toronto's successful bid for World Pride in 2009, significant efforts were undertaken to revitalize the village. Spearheaded by a local city councilor and the Church-Wellesley Business Improvement Association (BIA), these projects involved the commissioning of the Church Street murals to commemorate the village's LGBTQ history and the revitalization of Cawthra Park (since renamed Barbara Hall Park) (Nash & Gorman-Murray, forthcoming). These efforts were put in place to bolster the tourism trade that was expected with World Pride. Upon first glance, this motivation precludes considerations of a just city, but as Kanai and Kenttamaa-Squires (2015) argue, commemorative projects such as these work to (re)claim LGBTQ public spaces, an important aspect of the caring and repairing of public space. In this case, repairing public space involves 'queering' it, by creating murals on Church Street and an AIDS memorial in Cawthra Park, which 'reclaims' the spaces of the village to support local residents.

Care and repair in and of the gay village seemingly leads to more just public spaces. In certain ways it does, but this narrative is neither simple nor linear. In Sydney during the 1970s and 1980s, Oxford Street-based community organizations served primarily gay men, leading to the displacement of lesbians to other parts of the city (Jennings, 2015). Nash and Gorman-Murray (2015a) trace the movement of these women to the inner-west suburbs, where women's health, lesbian counseling and women's refuges were established. Mobility justice impels us to see how care services are moored in certain places for certain parts of the LGBTQ communities and prompts us to consider how care practices might be more equitably distributed across LGBTQ communities.

Public space and procedural justice

Low and Iveson's (2016) final lens examines procedural fairness around the constitution of public spaces. This focuses attention on how decisions about public spaces are made and the extent to which public spaces are "the object of genuinely democratic and inclusive public debate in the wider urban public sphere" (Low & Iveson, 2016, p. 21). While participants may be satisfied with the distributional outcomes of particular processes, "the favorability of an outcome was less crucial when the underlying allocation process was perceived as fair" (Low & Iveson, 2016, p. 21).

For the LGBT community, access to public spaces for socializing, or for more intimate encounters, has always been problematic. In the 1960s and 1970s in Toronto, the mainstream news media reported disapprovingly about gay men's activities in a number of public locations, including parks in late evening hours. In some cases, police set up 'sting' operations in the bathrooms of various bars to catch men engaging in sex acts. Police raids and liquor license inspections on gay venues, including clubs and bathhouses, often resulted in businesses closing and/or numerous arrests (Nash, 2005). Through formal administrative procedures in the 1980s and 1990s, the LGBTQ community was able to participate in decision-making processes about public space, meet in commercial venues and hold annual Pride celebrations without fear. Relations improved to the point where a contingent of Toronto police marched in uniform in Toronto's Pride parade from 2000 to 2016. However, in September–October 2016, police again set up a covert operation in a Toronto-area park in an attempt "to lure gay men and trans people to proposition them for sex" (Rieti, 2016, para. 5). Arguably, with new mobilities, and new technologies such as dating apps, some gay men are meeting others beyond inner-city locations in more suburban locations, where their presence is regarded as 'out of place'. The tensions between the police and the LGBTQ community show that access to public space, even when those locations are being used in the late evening and overnight, remains an issue.

At the same time, efforts to shore up the gay village as LGBTQ space continues, with the support (although not unanimous) of the City of Toronto municipal council. In anticipation of the World Gay Games being held in Toronto in June 2014, the city helped fund improvements in LGBTQ public spaces in the gay village. City support for the appropriation of public spaces as almost exclusively LGBTQ space helped ensure the legitimacy of LGBTQ public spaces, but this also might suggest that LGBTQ people have 'their own spaces' and therefore might be legitimately excluded from other locations.

Conclusions

Mobility is seen as a hallmark of the contemporary subject, and sexual and gender minorities are framed as quintessentially modern subjects 'on the move' through both social and geographical mobility (Puar, Rushbrook & Schein, 2003). Yet as Cresswell (2010) argues, mobility is political, and not every subject has the same motility or capacity for mobility. This is where questions of mobility justice – having the rights and the resources to move – intervene in discussions of new mobilities. While they sometimes have agency, queer subjects are equally impelled to move by specific sets of circumstances – to escape marginality and actualize self (Knopp, 2007) – although this is also embedded within economic and social constraints that close down as well as open up particular trajectories, paths and destinations. Mobility justice, then, needs to consider social differences, including those of sexuality and gender identity.

We suggest that the concept of mobility justice needs to consider the specific circumstances of queer mobilities at the urban scale. Low and Iveson's (2016) propositions for more just urban public spaces offer a useful lens for considering mobility justice in the contemporary city for sexual and gender minorities. Our application of these propositions to the circumstances of a particular social group – queer subjects in Toronto and Sydney – suggests that experiences of justice in the city remain linked to social differences and their intersections, which need to be taken into account if we are to advance mobility justice. As queer subjects appreciate, the city is not endlessly open to them. Some spaces are off limits, and movement is truncated due to impositions of social power and 'moral' geographies. This suggests we need to continue to consider the diverse dimensions of urban justice – redistribution, recognition, encounter, care and repair, and procedural justice – and how these might be actualized for sexual and gender minorities.

Note

1 The acronym LGBTQ refers to lesbian, gay, bisexual, trans and queer constituents. The first four identities are assumed to be relatively stable, and the basis for LGBT identity politics. Queer denotes those sexual and gender minority subjects who eschew LGBT identities and, indeed, often resist self-identification on the basis of sexual or gender performance.

References

Adey, P. (2006). 'If mobility is everything then it is nothing: Towards a relational politics of (im)mobilities'. *Mobilities*, 1(1), 75–94.

Bain, A., Podmore, J. & Rosenberg, R. (2017, April). 'Myopic homophobia: Print media (mis)representations of LGBTQ2 suburban lives in Canada's largest cities'. Paper presented at the American Association of Geographers Annual Meeting, Boston, MA.

Bell, D. & Binnie, J. (2000). *The sexual citizen: Queer politics and beyond*. Cambridge: Polity Press.

Binnie, J. (2004). 'Quartering sexualities: Gay villages and sexual citizenship'. In D. Bell & M. Jaynes (Eds.), *City of quarters: Urban villages in the contemporary city* (pp. 163–172). Aldershot: Ashgate.

Browne, K. (2007). 'A party with politics? (Re)making LGBTQ pride spaces in Dublin and Brighton'. *Social & Cultural Geography*, 8(1), 63–87.

Castells, M. (1983). *The city and the grassroots: A cross-cultural theory of urban social movements*. Berkeley, CA: University of California Press.

Collins, A. (2004). 'Sexual dissidence, enterprise and assimilation: Bedfellows in urban regeneration'. *Urban Studies*, 41(9), 1789–1806.

Collins, A. & Drinkwater, S. (2017). 'Fifty shades of gay: Social and technological change, urban deconcentration and niche enterprise'. *Urban Studies*, 54(3), 765–785.

Cook, N. & Butz, D. (2018). 'Gendered mobilities in the making: Moving from a pedestrian to vehicular mobility landscape in Shimshal, Pakistan'. *Social & Cultural Geography*, 19(5), 606–625.

Cook, N. & Butz, D. (2016). 'Mobility justice in the context of disaster'. *Mobilities*, 11 (3), 400–419.

Cresswell, T. (2010). 'Towards a politics of mobility'. *Environment & Planning D: Society & Space*, 28(1), 17–31.

Culver, G. (2017). 'Mobility and the making of the neoliberal "creative city": The streetcar as a creative city project?' *Journal of Transport Geography*, 58, 22–30.

Dufty-Jones, R. (2012). 'Moving home: Theorizing housing within a politics of mobility'. Housing, Theory & Society, 29(2), 207–222.

Fincher, R. & Iveson, K. (2008). *Planning and diversity in the city: Redistribution, recognition and encounter*. Basingstoke, UK: Palgrave.

Fraser, N. (1996). *Social justice in the age of identity politics: Redistribution, recognition and participation*. Tanner Lectures in Human Values. Stanford, CA: Stanford University. Retrieved from www.intelligenceispower.com/Important%20Emails%20Sent%20attachments/Social%20Justice%20in%20the%20Age%20of%20Identity%20Politics.pdf.

Freund, P. & Martin, G. (2007). 'Hyperautomobility, the social organization of space and health'. *Mobilities*, 2(1), 37–49.

Goggin, G. (2016). 'Disability and mobilities: Evening up social futures'. *Mobilities*, 11 (4), 533–541.

Gorman-Murray, A. & Nash, C. J. (2017). 'Transformations in LGBT consumption and leisure space in the neoliberal city'. *Urban Studies*, 54(3), 786–805.

Gorman-Murray, A. & Nash, C. J. (2014). 'Mobile places, relational spaces: Conceptualizing an historical geography of Sydney's LGBTQ neighborhoods'. *Environment & Planning D: Society & Space*, 32(4), 622–641.

Gorman-Murray, A. & Waitt, G. (2009). 'Queer-friendly neighborhoods: Interrogating social cohesion across sexual difference in two Australian neighborhoods'. *Environment & Planning A*, 41(12), 2855–2873.

Gorman-Murray, A., Waitt, G. & Gibson, C. (2012). 'Chilling out in cosmopolitan country: Urban/rural hybridity and the construction of Daylesford as a "lesbian and gay rural idyll"'. *Journal of Rural Studies*, 28(1), 69–79.

Jennings, R. (2015). *Unnamed desires: A Sydney lesbian history.* Clayton, AU: Monash University Publishing.

Jensen, A. (2011). 'Mobility, space and place: On the multiplicities of seeing mobility'. *Mobilities*, 6(2), 255–271.

Jensen, O. (2009). 'Flows of meaning, cultures of movement: Urban mobility as meaningful everyday life practice'. *Mobilities*, 4(1), 139–158.

Kanai, J. & Kenttamaa-Squires, K. (2015). 'Remaking South Beach: Metropolitan gayborhood trajectories under homonormative entrepreneurialism'. *Urban Geography*, 36(3), 385–402.

Kaufmann, V. (2002). *Rethinking mobility: Contemporary sociology.* Aldershot, UK: Ashgate.

Knopp, L. (2007). 'From lesbian to gay to queer geographies: Pasts, prospects and possibilities'. In K. Browne, J. Lim & G. Brown (Eds.), *Geographies of sexualities: Theory, practices & politics* (pp. 21–28). Farnham, UK: Ashgate.

Knopp, L. (1990). 'Some theoretical implications of gay involvement in an urban land market'. *Political Geography Quarterly*, 9(4), 337–352.

Leonard, W., Pitts, M., Mitchell, A., Lyons, A., Smith, A., Patel, S., Couch, M. & Barrett, A. (2012). *Private lives 2: The second national survey of the health and wellbeing of gay, lesbian, bisexual and transgender (GLBT) Australians.* Melbourne: The Australian Research Centre in Sex, Health & Society, La Trobe University. Retrieved from www.glhv.org.au/ les/PrivateLives2Report.pdf.

Low, S. & Iveson, K. (2016). 'Propositions for more just urban public space'. *City*, 20 (1), 10–31.

Low, S. & Smith, N. (2006). *The politics of public space.* London: Routledge.

Mackintosh, P. G. (2017). *Newspaper city: Toronto's street surfaces and the liberal press, 1860–1935.* Toronto: University of Toronto Press.

McCann, E. & Ward, K. (2011). *Mobile urbanism: Cities and policymaking in the global age.* Minneapolis, MN: University of Minnesota Press.

McCartan, A. (2017). *Glasgow's queer battleground.* Unpublished Master's thesis, Brock University, St. Catharines, ON.

Mitchell, D. (2003). *The right to the city: Social justice and the fight for public space.* London: The Guilford Press.

Montegary, L. & White, M. A. (Eds.). (2015). *Mobile desires: The politics and erotics of mobility justice.* Basingstoke: Palgrave Macmillan.

Mullen, C. & Marsden, G. (2016). 'Mobility justice in low carbon energy transitions'. *Energy Research & Social Science*, 18, 109–117.

Nash, C. J. (2013b). 'Queering neighborhoods: Politics and practice in Toronto'. *ACME: International E-Journal for Critical Geographies*, 12(2), 193–213.

Nash, C. J. (2013a). 'The age of the "post-mo"? Toronto's gay village and a new generation'. *Geoforum*, 49, 243–254.

Nash, C. J. (2006). 'Toronto's gay village (1969 to 1982): Plotting the politics of gay identity'. *Canadian Geographer/Le Géographe Canadien*, 50(1), 1–16.

Nash, C. J. (2005). 'Contesting identity: The struggle for gay identity in Toronto in the late 1970s'. *Gender, Place & Culture*, 12(1), 113–135.

Nash, C. J. & Gorman-Murray, A. (2015a). 'Lesbians in the city: Mobilities and relational geographies'. *Journal of Lesbian Studies*, 19(2), 173–191.

Nash, C. J. & Gorman-Murray, A. (2015b). 'Recovering the gay village: A comparative historical geography of urban change and planning in Toronto and Sydney'. *Historical Geography*, 43, 84–105.

Nash, C. J. & Gorman-Murray, A. (2014). 'LGBT neighborhoods and "new mobilities": Towards understanding transformations in sexual and gendered urban landscapes'. *International Journal of Urban & Regional Research*, 38(3), 756–772.

Nash, C. J. & Gorman-Murray, A. (forthcoming). 'LGBT place management: Representative politics and Toronto's gay village'. In M. Tremblay (Ed.), *LGBT and Electoral Politics in Canada*.

Nicholson, J. & Sheller, M. (2016). 'Race and the politics of mobility: Introduction'. *Transfers*, 6(1), 4–11.

Noack-Lundberg, K. (2012). *Queer women's experiences in public spaces*. Unpublished PhD thesis, La Trobe University, Melbourne, AU.

Puar, J., Rushbrook, D. & Schein, L. (2003). 'Guest editorial'. *Environment & Planning D: Society & Space*, 21(4), 383–387.

Reynolds, R. (2009) 'Endangered territory, endangered identity: Oxford Street and the dissipation of gay life'. *Journal of Australian Studies*, 33(1), 79–92.

Rieti, J. (2016, November 18). 'Toronto police should drop Project Marie charges, city and provincial politicians say'. *CBC News*. Retrieved from www.cbc.ca/news/canada/toronto/project-marie-reaction-1.3858328.

Ruting, B. (2008). 'Economic transformations of gay urban spaces: Revisiting Collins' evolutionary gay district model'. *Australian Geographer*, 39(3), 259–269.

Sheller, M. (2016). 'Uneven mobility futures: A Foucauldian approach'. *Mobilities*, 11 (1), 15–31.

Sheller, M. (2014). 'Mobility justice'. *Wi Journal of Mobile Culture*, 8(1). Retrieved from http://wi.mobilities.ca/wp-content/uploads/2014/09/wi_08_01_2014_Sheller.pdf.

Sheller, M. (2011). 'Sustainable mobility and mobility justice: Towards a twin transition'. In M. Grieco & J. Urry (Eds.), *Mobilities: New perspectives on transport and society* (pp. 289–304). London: Routledge.

Staeheli, L. & Mitchell, D. (2007). 'Locating the public in research and practice'. *Progress in Human Geography*, 31(6), 792–811.

Sullivan, A. (2005). 'The end of gay culture: Assimilation and its meanings'. *The New Republic*, 233(17), 16–21.

Tilley, C. & Hoad, N. (2017, October 11). 'A respectful debate: The same-sex marriage debate has been marred by hate speech, vandalism and bullying. Keep track of the incivilities'. *ABC News*. Retrieved from www.abc.net.au/news/2017-10-11/ssm-same-sex-marriage-respectful-debate-ugly-side/8996500.

Uteng, T. P. & Cresswell, T. (Eds.). (2016). *Gendered mobilities*. London: Routledge.

Valentine, G. (1995). 'Out and about: Geographies of lesbian landscapes'. *International Journal of Urban & Regional Research*, 19(1), 96–112.

Warner, T. (2002). *Never going back: A history of queer activism in Canada*. Toronto: University of Toronto Press.

Witten, K., McCreanor, T. & Kearns, R. (2003). 'The place of neighborhood in social cohesion: Insights from Massey, West Auckland'. *Urban Policy & Research*, 21(4), 321–338.

Wotherspoon, G. (2016). *Gay Sydney: A history*. Sydney, AU: NewSouth Press.

14 Mobility (in)justice, positionality and translocal development in Gojal, Pakistan

Andreas Benz

Introduction

The northern Pakistani region of Gojal is located in the province of Gilgit-Baltistan, in the midst of the Karakoram mountain range. Gojal has undergone exceptional transformation over the past decades, particularly in relation to improved levels of human development that are virtually unparalleled in other rural regions of the country (Benz, 2014a, 2014b). Mobility and migration have been important drivers and facilitating factors of educational achievement, poverty reduction, women's empowerment and off-farm employment by enabling Gojalis to access resources and opportunities outside the region as part of translocal livelihood strategies (Benz, 2016; Kreutzmann, 2012). However, Gojali households are differentially mobile. While most have benefited from mobility-related opportunities like higher education and professional employment in downcountry urban centers, some have not been in a position to appropriate such opportunities and develop migration strategies. Differential mobilities at the household level have produced significant and growing socioeconomic disparities within Gojali communities. Mobility-related inequalities have also developed within individual households as differently positioned members have differential access to mobility-based opportunities and social goods.

Because these new socioeconomic inequalities are produced in relation to the differential access to social goods that mobility affords, they raise concerns about social justice but also about mobility justice in Gojal. In this chapter, I scrutinize the particular interplay of social justice and mobility justice in the region, drawing on a case study of the neighboring villages of Hussaini and Passu, located in lower Gojal. I argue that in Gojal social justice is challenged by growing socioeconomic inequalities between the households of these communities and by the unequal distribution of opportunities, like education, employment and income, within households. I further argue that access to mobility and migration is a precondition for socioeconomic prosperity and access to these opportunities. Because unequal access to mobility is a major driver of social inequality, mobility justice is a precondition for social justice. Case study findings show that access to mobility and migration depends on social capital in the form of access to the 'right' kinship networks as well as on gendered social norms and social positionings within the household. I conclude that in Gojal, mobility justice and consequently social justice are governed mainly by social norms, household position and social capital as the major determinants for access to mobility.

My argument builds in three sections of the chapter. I first review recent conceptualizations of mobility justice and social justice to identify their constituting elements and determinants, and to shed light on their interrelations. In the same section I discuss the importance of social capital to the distributional dimension of mobility justice. Second, I describe Gojal's development path since the 1940s, highlighting the role played by mobility, migration and translocal social networks in that history. I use this historical sketch to explain the growing socioeconomic disparities within Hussaini and Passu villages with reference to households' differential access to network capital, which has shaped their participation in – or exclusion from – translocal livelihood strategies. Finally, I focus on dimensions of (in)justice at the household level, in which differently positioned members have differential access to mobility and its socioeconomic affordances based on gendered social norms.

Empirical data was collected during three months of field research in 2011 and 2012. Key data sources include migration histories of villagers from Hussaini and Passu that were collected through household-based village surveys of all 1,283 members (including temporary migrants) of the villages' 185 households, as well as more than 450 permanent outmigrants. These histories provide insight into villagers' migration biographies and educational and professional careers. I also draw from biographical oral history and focused narrative interviews conducted with former and current migrants, village elders, teachers and representatives of village organizations and social sector NGOs.

Conceptualizing mobility justice

Recently, advances have been made within mobility studies to theorize the relationship between social justice and spatial mobility (e.g., Cook & Butz, 2016; Montegary & White, 2015; Sheller, 2015a, 2015b). Notions of the "politics of mobility" (Adey, 2006; Cresswell, 2010) and "motility" (Elliott & Urry, 2010; Kesselring, 2006; Sheller, 2014), understood as "mobility capital" (Kaufmann, Bergmann & Joye, 2004) or the "capacity for movement ... under conditions of one's choosing" (Cook & Butz, 2016, p. 400), have also been useful in considering the unequal distribution of access to mobility as both an outcome of existing power asymmetries and a mechanism of reproducing socioeconomic inequalities. In these strands of research, differential mobilities have systematically been linked to social exclusion. The unequal distribution of mobility capital may enable, facilitate or speed up the mobility of some, while at the same time slowing down, hampering or inhibiting the mobility of others, thus constraining their ability to actively participate in social and political life and to access economic opportunities (Sheller, 2014, p. 798). According to Cook and Butz (2016, p. 401), critical mobility studies thus center around questions related to social justice: "Who is able to access and appropriate mobility capital?" and "How broadly are capabilities of movement extended throughout a social system?" These questions are also central to issues of mobility justice.

Cook and Butz (2016) develop the concept of mobility justice in relation to social justice. For this purpose they draw on Young's (1990) notion of social justice, which considers not only the equitable distribution of social goods and harms, but also domination, the institutional and structural context of distribution, including rule-setting procedures and the opportunities of individuals to participate in decision-making processes. When applied to mobility, this concept of justice suggests that attention needs to be paid to inequalities in the "uneven distribution of capacities and competencies [for mobility] in relation to the physical, social and political affordances for movement" (Sheller, 2014, p. 797), but also the (un)equal participation of individuals in the governance of mobility systems.

While Cook and Butz (2016) put the notion of domination at the center of their mobility justice approach, Sheller (2008, p. 31) instead focuses on the notion of freedom. She differentiates three senses or meanings of mobility justice: (a) the degree of "personal freedom of mobility", which is related to the distribution of mobility rights to individuals and the actual provision of access to mobility; (b) the "sovereignal freedom of mobility", meaning the power of individuals to control others' personal freedom of mobility; and (c) the degree of "civic freedom of mobility", implying the power to determine mobility systems and rights in a society, as well as possibilities for individuals to participate in this decision-making. Sheller (2008, p. 28) points out that certain interdependencies exist among these different meanings. For example, realizing some people's personal freedom of mobility may imply the exercise of sovereignal freedom of mobility by restricting the personal freedom of mobility of others, potentially creating a situation of mobility injustice; some mobilize, others are demobilized, and both processes are closely interlinked.

These approaches to mobility justice share common elements. For instance, Sheller's personal and sovereignal freedoms of mobility are commensurate with the distributional meaning of mobility justice in Cook and Butz's work, while the civic freedom of mobility speaks to mobility justice's institutional and structural contexts. In both approaches, the distribution of mobility capacities is shaped by the institutional and structural context in question. This context includes social capital in the form of social networks, gendered social norms and the institution of the household, all of which are central to the mobility context of Gojal. While I later reflect on gendered norms and the household, in what follows I outline what I mean by social capital and social networks.

Portes (1998, p. 6) defines social capital as the "ability to secure benefits through membership in networks and other social structures", which can range from family and kinship to ethnic and religious networks. He delineates three basic effects of social capital. First, it can serve as a "source of social control" as found in structures of bounded solidarity and enforceable trust (Portes, 1998, p. 9). The social capital created by tight community networks may be useful to parents, teachers and police authorities as they seek to maintain discipline and promote compliance among those under their charge. Second, it may serve as a "source of parental and kin support" (Portes, 1998, p. 10). And third, social capital constitutes a "source of benefits through extrafamilial networks" (Portes, 1998, p. 9), such as communal networks based on shared imaginations and constructions of identities, ethnicities, origin, descent or religion (Portes, 1998, p. 13).

Whether, and to what degree, an individual enjoys the benefits of social capital in kinship and extrafamilial networks largely depends on their membership and position in these networks (Portes, 1998, p. 13). Unequal access to networks and different positions within networks leads to unequal access and distribution of assets and benefits. These distributive effects of social capital have often served as an explanation for social stratification and unequal access to social goods and opportunities, such as education, employment and social mobility (Bourdieu, 1977; Portes, 1998, p. 13). Equally, I argue, social capital is decisive in distributing mobility capacities by providing or withholding the means and opportunities for mobility, thus enhancing or restraining personal freedom of movement. Therefore, the effects of social capital as a source of kinship or extrafamilial support tie in with the distributional meaning of mobility justice. In turn, the social control effect of social capital ties in with the institutional and structural context of mobility justice. It facilitates enforcement of rules and norms and, thus, may curtail individual freedom of movement.

In the context of migration, social capital created by migration networks plays a central role in enabling and facilitating mobility (Castles, 2010; Faist, 1998; Massey, 1990). Consequently, Ernste, Martens and Schapendonk (2012, p. 510) consider migrants' social networks as an institutional structuring force that enables the movement of people, goods, capital and information. Given the selective membership and differential positions in migration networks, these structures enable and facilitate migration for some but exclude others. The uneven individual endowment with social capital is a decisive factor in determining unequal mobility capacities and access to mobility. Consequently, any inquiry into mobility justice should pay close attention to the role of differential access to and positionalities within social networks.

The following case study focuses on the role played by social networks in distributing mobility capital to individuals. I argue that unequal access to translocal social networks is a decisive aspect of mobility (in)justice in Gojal. Who is able to access mobility is predicated in this context on membership in powerful translocal networks. Over the last several decades, many Gojali families have established social networks and moorings in a range of down-country cities and, thus, have been able to enhance socioeconomic opportunities for their members. Other families, however, have not been in a position to spatially diversify their livelihoods through translocal networks. I explain the reasons for these differential mobility paths and unequal distributions of mobility capital and draw out some of their implications. I begin by describing the process of mobility expansion in Gojal since the 1940s.

Translocal development in Gojal

Gojal is home to about 20,000 villagers of Ismaili faith and Wakhi ethnolinguistic heritage. Historically, Gojalis have experienced extreme poverty, frequent famine and pervasive illiteracy (Kreutzmann, 1989, p. 162, 1996, p. 289; Malik & Piracha, 2006, p. 360). But since the late 1940s, the region has realized impressive advancements in people's wellbeing and is currently well known for its impressive levels of human development (Kreutzmann, 1996; World Bank, 2011). Educational achievement and gender equality, for example, have reached levels virtually unparalleled in other rural areas of Pakistan (Benz, 2014a, p. 99; Felmy, 2006). Gojalis' mobility and migration strategies were key to enabling these developments (Butz & Cook, 2011; Kreutzmann, 1991, 1993, 2012).

Migration from Gojal began in the 1940s when the region was integrated into the Pakistani state and rigid travel restrictions were eased (Kreutzmann, 1996, p. 289; Sökefeld, 1997, p. 87). Massive road construction projects, particularly the Karakoram Highway, improved accessibility, thereby fostering mobility, increasing exchange flows and creating new livelihood opportunities such as cash crops, trade and tourism (Allen, 1989; Kamal & Nasir, 1998; Kreutzmann, 1991; Malik & Piracha, 2006). Increasing numbers of young men who left in the 1950s and 1960s to serve in the army or as unskilled laborers remitted earnings to their families.

At the onset of outmigration, local communities were characterized by a high degree of socioeconomic homogeneity in terms of livelihoods, income, occupations and landholdings (Cook & Butz, 2016, p. 408; Wood & Malik, 2006). As new livelihood opportunities became accessible, social inequalities increased. Different households incorporated new mobility options into their livelihood strategies in differentiated ways, initiating a long-term process of growing inequality in terms of unequal access to non-local resources, such as education and jobs.

Initially, Gojali migrants focused on Karachi, where they found support from the affluent Ismaili community, in the form of jobs in their factories, hotel chains and other businesses, and in other Ismaili networks and institutions (Kreutzmann, 1989, p. 192, 1996, p. 35). This social capital in extrafamilial networks was based on solidarity among members of the same identity group without prior personal acquaintance. Quickly, Gojalis and other Ismaili migrants from northern Pakistan built their own support networks in Karachi to facilitate new migration, and they provided assistance for new arrivals. They organized food, dwelling and jobs for newcomers, provided contacts and information and introduced them to the local Ismaili community. These forms of solidarity were based on symbolic ties of religion, ethnicity and region, but increasingly also on agnatic kinship relations, indicating a shift in the effects of social capital from support through extrafamilial networks to forms of kinship-based support. The next generation of Gojali migrants spread beyond Karachi to other lowland cities, continuously expanding their migration and support networks. The earnings of soldiers and unskilled laborers of the first migrant generation were used within translocal kinship networks to enable other family members to acquire higher education in urban centers. After graduation, this second generation often built highly skilled professional careers, in turn supporting subsequent generations' migration for higher education. Since the 1980s, increasing numbers of women have joined this migration pathway.

The coupling of one generation's success with their readiness to support a second wave of migration triggered an upward spiral of rising education and income levels among Gojalis. Thanks to well-established kinship solidarity and resource redistribution systems, many family members benefited whenever an individual secured a professional position and salary. Dominant values of sharing, solidarity and reciprocity prevent individuals from hoarding resources and encourage the redistribution of money and provision of accommodation, food, contacts and assistance in different locations within kinship networks. These support systems, which constitute important aspects of the regional institutional and structural contexts of mobility justice, are not restricted to the household or nuclear family, but usually include larger family networks. However, they always exclusively run along lines of agnatic relations. At marriage, women leave the parental household to join their in-laws' family networks, which truncates virtually all support flows to and from natal families. The basic unit of support, therefore, is patrilineal branches within kinship networks.

This extensive translocal solidarity and family support has provided new opportunities for higher education and professional employment outside Gojal for many villagers. My survey data show that on average households in Passu and Hussaini spend about one-third of their disposable income on education, which has allowed Gojali youth to acquire the requisite educational credentials, knowledge, skills and experience to land remunerative jobs in the private and government sectors or to establish their own enterprises. Most migrants have returned – at least temporarily – to their home regions to serve their communities, as teachers, medical specialists, engineers, financial experts, development consultants and entrepreneurs, and to take on responsibility for mobilizing their communities and instigating social change. But while absent, they retain close ties with their families and communities, and they actively contribute to their development through financial and social remittances.

Migration and improved wellbeing are closely intertwined in the case of Gojal (Kreutzmann, 1989, pp. 180–195, 1993, 2012). Consequently, migration has been identified as a "key livelihood option" in this area (Wood & Malik, 2006, p. 73). It is linked to positive effects on education, income, investments and living standards for most households, but not for all, and not in the same way. But this positive general trend demands further differentiation. Not all households participated in the early phase of outmigration, which was dominated by military and labor migrants. This differentiation had significant implications for households' ability to engage in later phases of migration, particularly for education. In most cases, only those households that participated in early outmigration were in a position to send children out for education, because only they had the requisite financial capital (remittances and savings of early migrants) and access to translocal networks (family members living in other places that could provide *in situ* support). Consequently, differences in the potential of certain kinship networks to support migration developed at this early stage. These differences multiplied in the course of intensifying outmigration from Gojal, because kinship networks with a high number of migrants proved to be more capable of supporting the next generation of migrants, while kinship networks with few or no migrants struggled to participate in migration activities due to lack of remittances and weak translocal support networks. Access to the 'right' networks, then, significantly enhances the mobility capital of a household or individual, which again is a precondition for access to social goods, such as education. Because unequal access to networks translates into unequal education levels, which then determines opportunities for professional careers and income levels, social capital created in kinship networks becomes an issue of mobility justice in its distributional meaning.

Today, migration and mobility are significant aspects of Gojalis' lives, as demonstrated by the region's rate of labor outmigration, which is well above the average of rural communities in other parts of Pakistan (World Bank, 2011, p. 20). According to my 2012 survey, 30% and 41% of the male workforce in Hussaini and Passu, respectively, have migrated to cities outside Gilgit-Baltistan. Among youth aged 15 to 24, migration rates were even higher: 63% in Hussaini and 77% in Passu. In total, 69% of households in Hussaini and 79% in Passu had at least one family member living outside Gojal at the time. Another 21% of households in both Hussaini and Passu that had no current member living outside included at least one returned migrant, indicating their former participation in migration strategies. All Passu households had at one time employed migration as part of their livelihood strategies. In contrast, 8 out of 84 households in Hussaini had yet to have a member migrate. Consequently, these households are characterized by comparatively low adult education levels, low monetary household income, lack of members in formal employment, high dependency on agriculture and occasional laboring, and by a four times higher prevalence of multidimensional poverty. Subsistence agriculture plays a more important role in their livelihoods, but they have fewer livestock and lower cash-crop income compared to households that have used migration strategies. In addition, they have very few, if any, migrants among their agnatic kin, which limits their translocal links, opportunities for support at potential migration destinations and translocal assistance within their family network. These mobility-poor Gojalis are simultaneously economically poor.

Similarly, my data shows a clear correlation between the level of migration activity and levels of education, monetary income and professional employment in those Hussaini and Passu households that have employed migration strategies This correlation underscores the importance of access to mobility as a precondition for tapping new external livelihood opportunities and for overcoming local constraints and limitations with respect to education, employment and income.

Mobility justice, gendered social norms and household positionality

So far, I have demonstrated that the distribution of mobility capabilities is uneven among households in Hussaini and Passu. Here I explain their inequitable distribution *within* households. Given that individual opportunities for mobility are a precondition for access to education and employment, and subsequently to economic prosperity, the unequal distribution of mobility capital among household members decisively affects mobility justice and social justice. Following Sen's (1990, p. 123) conception of the household as an arena of "co-operative conflicts", differently positioned members, enmeshed in asymmetric power relations, compete for scarce household resources. Power relations pertaining to gender, age and marital status are frequently exercised in ways that lead to the unequal distribution of resources and opportunities.

The most senior male usually occupies the most powerful position in a Gojali household, followed by junior male members. Women occupy subordinate positions. A woman's status improves with age and the number of children she bears, particularly sons. The weakest position in a household is that of a daughter-in-law before giving birth to her first child (Felmy, 1996, p. 20). These differently positioned household members experience differential access to mobility as a social good, just as they do with respect to other resources and opportunities, resulting in differential mobility capabilities. Decisions about the distribution of resources and opportunities are generally made by the head of the household, who sometimes consults with other household members.

Household heads often draw on dominant gender norms when making decisions about their children's educational opportunities. Boys' education has been prioritized over girls' education since the first boys-only schools were established in Gojal in the late 1940s. Not until the 1970s, with external incitement by the Ismailis' spiritual leader, the Aga Khan, did the first girls' schools open, encouraging parents to dedicate household resources to their daughters' education. Today, most households value education for their sons and daughters equally. However, many parents continue to argue that significant investments in their daughters' higher education do not make economic sense because at marriage they leave the parental household to become part of a new family. Further investments in sons' education, in contrast, have long-lasting benefits for parents; sons generally assume responsibilities for the care of their parents in old age.

This reluctance to invest in girls' higher education becomes particularly apparent when households make decisions about student outmigration. While overall educational achievement is fairly equal in the young generation, a pronounced gender gap exists in patterns of education migration to places outside Gilgit-Baltistan. Young men have many opportunities to acquire higher education in Pakistani lowland cities, and they tend to receive a disproportionate share of household mobility investments. This differential intrahousehold resource distribution is influenced by expectations regarding the rates of return on investments (e.g., future employment and income opportunities), but also by gendered norms that shape the gendered division of household labor. Women's extensive list of household tasks and farm responsibilities keeps them close to home and leaves little time for travel. Gender-related differences in mobility opportunities and mobility costs pertain to travel constraints for women and the gendered concept of honor that restricts young women's interactions with non-kin men and the range of appropriate living arrangements away from home (Cook & Butz, 2018; Gioli et al., 2014; Gratz, 1998). Women are expected not to travel alone beyond the region, necessitating the accompaniment of a male relative or group of women. Female student migrants cannot rent a flat alone or share an apartment with fellow female students; rather, they are required to join the households of male relatives at the migration destination, in which they undertake chores that often hamper their educational progress. The only alternative is living in a student hostel, but costs are extremely high, which is another significant disincentive for girls' education.

The need for special travel arrangements and living accommodations leads to considerably higher mobility costs for female student migration. These costs severely constrain girls' access to higher education outside the region, thereby limiting their access to translocal livelihood opportunities. This form of mobility injustice interrelates with social injustice in the reproduction of gendered norms, mobilities and power imbalances. However, when compared to other regions of Pakistan, levels of education and professional employment for Gojali women are extraordinarily high and rising.

Given the comparatively higher costs and efforts associated with female student migration, their mobility opportunities largely depend on membership in the 'right' kinship networks. Only well-established translocal networks provide sufficient monetary resources from urban professional employment to cover educational costs and hostel fees if family living accommodations are not available. Female student migration, therefore, gained momentum only in the 1990s, after generations of pioneering male migrants established these translocal networks (Benz, 2016, p. 149). Given prevailing gender norms, pioneering female migration would have been unthinkable.

Currently, many men are engaged in outside income-generating activities and educational endeavors, while most women and the elderly stay put in their villages, compensating for absent men by taking over their agricultural work, household chores and social responsibilities in addition to their own reproductive and subsistence work (Cook & Butz, 2018). Mobile men thus exercise 'sovereign mobility freedom', which negatively affects other family members' freedom of mobility, leading to mobility injustice within the household. It is precisely the immobilization of women and elderly people that enables men to be mobile. Male and female mobilities are tightly connected and interdependent, forming a system of relative (im)mobilities (Cook & Butz, 2018). Those who stay put fulfill important tasks – caring for children, the sick and the elderly, tending cattle, cultivating gardens and fields, maintaining houses and material property – to the benefit of absent men, households, families and communities.

Due to gender norms and related mobility restrictions, then, women who might hope to pursue university studies outside the community are often pushed by their families and in-laws into the 'traditional' role of mother and housewife, confined to the realm of the household and subsistence production. Even many of those who do achieve an advanced education take up this role (or are forced into it), with only a few entering professional remunerated employment. Among 20- to 40-year-old women in Hussaini and Passu (excluding students), only 23% were pursuing remunerated employment, compared to 77% of their male counterparts.

Conclusions

This Gojali case study points to the important role of translocal opportunity structures, mobility and networks in improving livelihoods and social wellbeing. I demonstrated how the diversification of livelihood strategies through new opportunities and income-generating activities pursued in places outside Gojal has led to unprecedented socioeconomic advancement and prosperity. A precondition for such translocal livelihoods is mobility, which in turn links the ability to access social opportunities as a precondition of social and economic wellbeing with the ability to move. I also showed that unequal access to mobility at the household and individual levels has resulted in growing socioeconomic disparities and gendered discrimination. Therefore, questions of social (in)justice in the context of translocal opportunity structures are bound up with questions of mobility (in)justice. Access to mobility depends on social capital in the form of translocal networks as well as on gendered social norms and social positionings within the household. In the following I draw conclusions about the role of social networks before turning to the importance of norms and positionings.

An analysis of the history of outmigration from Gojal since the 1940s reveals how, at the household level, access to mobility has been governed by social capital in the form of access to potent translocal networks that provide support and assistance to migrants. Depending on the translocal assets available in these networks (e.g., remittances, on-site support such as providing contacts, information, free board and lodging), differential degrees of support and facilitation for mobility and migration can be provided. Unequal network access has resulted in unequal mobility capital and consequently unequal access to new livelihood opportunities and economic prosperity. While many

Gojali households can benefit from new translocal livelihood opportunities to improve their wellbeing, a minority of disadvantaged, mobility-poor households are increasingly marginalized from the overall trends of educational expansion, professionalization and improved wellbeing, resulting in a growing socioeconomic gap among households in Gojal's mountain communities. This situation underlines the central importance of social capital to the distributional dimension of mobility justice.

At the intra-household level, access to mobility is shaped by gendered norms and social positionings, as is shown through an analysis of the household as an arena of cooperative conflict, in which powerful household members distribute available assets and assign livelihood opportunities among household members, resulting in mobility injustices along lines of gender, age and marital status. Some household members are forced to refrain from mobility, and those staying put bear the costs of others' mobility.

I argued that the current situation of deepening social inequalities and growing social injustice in the region is largely attributable to the situation of mobility injustice derived from unequal network access, gendered norms, household positionings and gendered mobilities. Social justice would imply equal access to opportunities for all people of the region irrespective of their gender, family background and socioeconomic position. In a situation where social prosperity and advancement depends on access to mobility, creating and securing social justice depends on achieving mobility justice. Attempts to prevent a further widening of the social gap need to take into account the root causes of mobility injustice and pay attention to marginalized households' lack of mobility capital and the unequal distribution of mobility capital within households. This strategy of overcoming social injustice by tackling mobility injustice could comprise, for instance, targeted scholarship programs for student migration, improved education facilities within the region, provision of affordable hostel capacities at migration destinations and subsidized and reliable public transport systems. These measures should focus primarily on disadvantaged groups such as girls and women as well as members of socioeconomically poor and less translocally connected households. In this way, the impressive socioeconomic advancement that has resulted through migration as "the main livelihood story" of the last few decades (Wood & Malik, 2006, p. 73) could eventually become a success story including all Gojalis.

References

Adey, P. (2006). 'If mobility is everything then it is nothing: Towards a relational politics of (im)mobilities'. *Mobilities*, 1(1), 75–94.

Allen, N. (1989). 'Kashgar to Islamabad: The impact of the Karakorum Highway on mountain society and habitat'. *Scottish Geographical Magazine*, 105(3), 130–141.

Benz, A. (2016). 'Framing modernization interventions: Re-assessing the role of migration and translocality in sustainable mountain development in Gilgit-Baltistan, Pakistan'. *Mountain Research & Development*, 36(2), 141–152.

Benz, A. (2014a). *Education for development in northern Pakistan: Opportunities and constraints for rural households.* Karachi, PK: Oxford University Press.

Benz, A. (2014b). 'Mobility, multilocality and translocal development: Changing livelihoods in the Karakoram'. *Geographica Helvetica,* 69(4), 259–270.

Bourdieu, P. (1977). *Outline of a theory of practice.* Cambridge: Cambridge University Press.

Butz, D. & Cook, N. (2011). 'Accessibility interrupted: The Shimshal Road, Gilgit-Baltistan, Pakistan'. *Canadian Geographer,* 55(3), 354–364.

Castles, S. (2010). 'Understanding global migration: A social transformation perspective'. *Journal of Ethnic & Migration Studies,* 36(10), 1565–1586.

Cook, N. & Butz, D. (2018). 'Gendered mobilities in the making: Moving from a pedestrian to vehicular mobility landscape in Shimshal, Pakistan'. *Social & Cultural Geography,* 19(5), 606–625.

Cook, N. & Butz, D. (2016). 'Mobility justice in the context of disaster'. *Mobilities,* 11 (3), 400–419.

Cresswell, T. (2010). 'Towards a politics of mobility'. *Environment & Planning D: Society & Space,* 28(1), 17–31.

Elliott, A. & UrryJ. (2010). *Mobile lives.* London: Routledge.

Ernste, H., Martens, K. & Schapendonk, J. (2012). 'The design, experience and justice of mobility'. *Tijdschrift voor Economische en Sociale Geografie,* 103(5), 509–515.

Faist, T. (1998). 'Transnational social spaces out of international migration: Evolution, significance and future prospects'. *European Journal of Sociology,* 39(2), 213–247.

Felmy, S. (2006). 'Transfer of education to the mountains'. In H. Kreutzmann (Ed.), *Karakoram in transition: Culture, development and ecology in the Hunza Valley* (pp. 370–381). Karachi, PK: Oxford University Press.

Felmy, S. (1996). *The voice of the nightingale: A personal account of the Wahki culture in Hunza.* Karachi, PK: Oxford University Press.

Gioli, G., Khan, T., Bisht, S. & Scheffran, J. (2014). 'Migration as an adaptation strategy and its gendered implications: A case study from the Upper Indus Basin'. *Mountain Research & Development,* 34(3), 255–265.

Gratz, K. (1998). 'Walking on women's paths in Gilgit: Gendered space, boundaries and boundary crossing'. In I. Stellrecht (Ed.), *Karakorum – Hindukush – Himalaya: Dynamics of Change, Part II* (pp. 489–508). Köln: Rüdiger Köppe Verlag.

Kamal, P. & Nasir, M. (1998). 'The impact of the Karakorum Highway on the landuse of the Northern Areas'. In I. Stellrecht (Ed.), *Karakorum – Hindukush –Himalaya: Dynamics of change, Part I* (pp. 303–338). Köln: Rüdiger Köppe Verlag.

Kaufmann, V., Bergmann, M. & Joye, D. (2004). 'Motility: Mobility as capital'. *International Journal of Urban & Regional Research,* 28(4), 745–756.

Kesselring, S. (2006). 'Pioneering mobilities: New patterns of movement and motility in a mobile world'. *Environment & Planning A,* 38(2), 269–279.

Kreutzmann, H. (2012). 'After the flood: Mobility as an adaptation strategy in high mountain oases. The case of Pasu in Gojal, Hunza Valley, Karakoram'. *Die Erde,* 143(1–2), 49–73.

Kreutzmann, H. (1996). *Ethnizität im Entwicklungsprozess: Die Wakhi in Hochasien.* Berlin, DE: Reimer.

Kreutzmann, H. (1993). 'Challenge and response in the Karakoram: Socioeconomic transformation in Hunza, Northern Areas, Pakistan'. *Mountain Research & Development,* 13(1), 19–39.

Kreutzmann, H. (1991). 'The Karakoram Highway: Impact of road construction on mountain societies'. *Modern Asian Studies*, 25(4), 711–736.

Kreutzmann, H. (1989). *Hunza: Ländliche Entwicklung im Karakorum*. Berlin, DE: Reimer.

Malik, A. & Piracha, M. (2006). 'Economic transition in Hunza and Nagar Valleys'. In H. Kreutzmann (Ed.), *Karakoram in transition: Culture, development and ecology in the Hunza Valley* (pp. 359–369). Karachi, PK: Oxford University Press.

Massey, D. (1990). 'Social structure, household strategies, and the cumulative causation of migration'. *Population Index*, 56(1), 3–26.

Montegary, L. & White, M. A. (2015). 'The politics and erotics of mobility justice: An introduction'. In L. Montegary & M. A. White (Eds.), *Mobile desires: The politics and erotics of mobility justice* (pp. 1–14). London: Palgrave Macmillan.

Portes, A. (1998). 'Social capital: Its origins and applications in modern sociology'. *Annual Review of Sociology*, 24, 1–24.

Sen, A. (1990). 'Gender and co-operative conflicts'. In I. Tinker (Ed.), *Persistent inequalities: Women and development* (pp. 123–149). Oxford: Oxford University Press.

Sheller, M. (2015a). 'Uneven mobility futures: A Foucauldian approach'. *Mobilities*, 11(1), 15–31.

Sheller, M. (2015b). 'Racialized mobility transitions in Philadelphia: Urban sustainability and the problem of transport inequality'. *City & Society*, 27(1), 70–91.

Sheller, M. (2014). 'The new mobilities paradigm for a live sociology'. *Current Sociology Review*, 62(6), 789–811.

Sheller, M. (2008). 'Mobility, freedom and public space'. In S. Bergmann & T. Sager (Eds.), *The ethics of mobilities: Rethinking place, exclusion, freedom and environment* (pp. 25–38). Aldershot, UK: Ashgate.

Sökefeld, M. (1997). 'Migration and society in Gilgit, Northern Areas of Pakistan'. *Anthropos*, 92(1–3), 83–90.

Wood, G. & Malik, A. (2006). 'Sustaining livelihoods and overcoming insecurity'. In G. Wood, A. Malik & S. Sagheer (Eds.), *Valleys in transition: Twenty years of AKRSP's experience in northern Pakistan* (pp. 54–119). Karachi, PK: Oxford University Press.

WorldBank. (2011). *Gilgit-Baltistan economic report: Broadening the transformation*. Islamabad, PK: World Bank.

Young, I. M. (1990). *Justice and the politics of difference*. Princeton, NJ: Princeton University Press.

Justice and more-than-human mobilities

15 Mobility, animals and the virtue of justice

Fredrik Karlsson

Introduction

In this chapter, I argue from a non-representationalist point of view that the capability of mobility – the practical freedom to move – serves to develop a virtue of justice. Also, I conclude that the scope of mobility justice is open-ended concerning the inclusion of nonhuman agents. Moving accumulates memories and adaptations of successful action in different material and social settings, thereby increasing the ability of agents to handle the complexity of environments. The components necessary for acts of justice – autonomy and intentional action – are consequences of being able increasingly to handle the complexity of the material world. This ability is not necessarily unique to human beings, and, therefore, nor are genuinely just acts.

Non-representational theory has its origins in early pragmatism and acquired some inspiration from phenomenology. The world, it claims, consists of complex material relations that in part constitute phenomena that human beings experience as social, moral, spiritual, etc. There is nothing above and beyond materiality, although materiality is above and beyond strictly quantifiable parameters (Thrift, 2008, pp. 5–6). These relations are most obvious through shared experiences of the world that imbue language in an ongoing historical process. The mobility discourse I adhere to here is strongly influenced by non-representationalist thinking, which views mobility as the main embodied thrust through which shared experiences of material resistance and agential possibilities are made. Through mobility we may be exposed to new places, technologies, species and people, expanding the relational network of which we are a part and, in turn, bringing new opportunities for successful communication and common meaning-making (that may be oppressive) (Adey, 2017; McNay, 2004).

Associating mobility with the political-philosophical etiquette of 'capability' reflects such a material and relational stance, and situates mobility as a social justice concern. Amartya Sen (2001) developed the concept of capability in the 1970s to identify aspects of human flourishing, which involve the opportunity to exercise one's practical freedoms. Capabilities are actual opportunities that institutional arrangements (like infrastructure) make possible. They are also intrinsic to dignity; capabilities must be upheld in an equal manner for all to establish dignity and, thereby, social justice.

Aspects of Martha Nussbaum's view on capabilities are useful for non-representational theorizing. First, she situates capabilities in a neo-Aristotelian framework, stressing "the animal and material underpinnings of human freedom" as well as people's natural reciprocity that extends beyond family ties (Nussbaum, 2006, p. 88). Second, Nussbaum (2006, pp. 76–78, 159–160) argues that the rationale upon which society – and, thus, social justice – should be based is driven by bodily needs (formalized into a list of capabilities), and she uses Marx to argue that the purpose of capabilities is to make possible an "irreducible plurality of opportunities for life activity" (p. 167). It is not hard to see that such agency also would demand the capability for mobility, although Nussbaum does not include it in her list of capabilities. Third, her close association of the issue of capabilities and social justice with materiality expands the notion of capabilities to include sentient, nonhuman animals. Because social justice is based on bodily needs that pertain to human and nonhuman animals, analogous lists of capabilities should be shaped for animals, and these should be respected as a matter of social justice (Nussbaum, 2006, pp. 326, 392).

Calling mobility a capability, therefore, adheres to the view of mobility as an existential and benign condition of being in the relational complexity of the material world. It is a matter of social justice to acknowledge that mobility is necessary for beings to flourish (Bergmann & Sager, 2008, p. 3; Kronlid, 2016; Merriman, 2012). That would also include nonhuman agents.

Mobility justice *with* animals

The mobilities literature has begun to draw on social justice theory to frame uneven mobilities as a justice issue for human populations. This debate on social justice within the mobility discourse has reached a stage where it has become possible to prepare for, identify and/or promote sound normative guidelines. Familiar ethical approaches have appeared: the ideal is equal human worth, the object of critique is oppressive power relations, and the solution is emancipation (Adey, 2017; Cresswell, 2010; Martens, 2012; Sager, 2018, pp. 80–82). This exploratory work also has increased our understanding of the significance of mobility as a capability (Kronlid, 2016).

Nevertheless, I follow up on other trains of non-representationalist thought, even if they are not explicitly concerned with mobility. The reason is that the kind of normative work referred to above is concerned with problems (traffic, migration, etc.) involving human agents within political settings defined by human standards (*realpolitik*), which risks perpetuating the anthropocentric fallacy. Using this framework, at best we can, by analogy, adapt lists of human capabilities, or other anthropocentric norms, to other species (Nussbaum, 2006, pp. 392–401). However, as I have previously argued (Karlsson, 2012), such an approach typically reproduces anthropomorphic claims about the relatively low significance of other species. In contrast, Cary Wolfe (2010) and Nigel Thrift (2008) theorize agency and being as human and more-than-human phenomena. I draw on their and others' work to say something about the meaning of social justice when nonhuman agents are considered in a world of constantly flowing dynamics.

The stance on ethical theory

An important domain of change that preoccupies non-representational theorists is perception, a term that is used to speak of materiality beyond data. Perception is vital in enabling humans to orient themselves in an ever-changing world. Conscious decisions about how to act are made precognitively based on habitual perception, including affect (Thrift, 2008, pp. 6–7, 116–117). Thrift (2008, pp. 179–187) argues that certain affects have become part of our biologies through evolutionary processes, but also that perceptive habits are shaped in, and often made coarse by, practices relying on cultural codes, political processes, media and technological advances. Agency is heavily influenced by perception, which is a kind of interactional intelligence that develops in social practice. Wolfe also uses perception to link individual agency with systemic processes. He uses Niklas Luhmann's systems theory to conclude that the environment, which harbors closed, self-sustaining systems (organisms), forcefully channels the agency of those systems to perceive meaning in the world (Wolfe, 2010, pp. 18–19). Therefore, he develops a close association between exercising agency and developing perception. As I clarify below, this relationship between agency and perception affects how moral agency and social justice in a more-than-human world are understood.

Within ethical theory, the ability to saliently perceive moral aspects of a particular situation is most often tacitly accepted as a necessity (how could anyone do right without perceiving what is wrong?). In some cases, theorists reframe the issue of perception as one of empathy or other emotions (Donovan, 1996; Hume, 1998[1751]; Noddings, 2003; Nussbaum, 2003, pp. 19–88). Within a non-representationalist methodology, affect is indeed viewed as a significant expression of the relational processes involved in sentient beings' struggles to make sense of their environment. Perception, however, is not viewed as starting and ending with an empathetic individual.

The work of Iris Murdoch and Simone Weil is useful in associating a non-representationalist understanding of perception with ethical theory. This might be surprising as both are Platonists. Murdoch (2001, pp. 33–49) suggests that moral perception involves sensing traces of transcended, objectively existing values, indirect clues that are experienced as beauty. Weil (1972) argues that attention relates to objective states such that attentive love emanates from God. Human beings, attempting to perceive the world through love, rely on the inclusion of attentive love in a theocentric, self-referential loop (Murdoch, 2001, pp. 58, 82; Weil, 1972, p. 57). These insights are useful for non-representational theory because they do not limit moral perception to a human skill *per se*. Rather, it is a consequence of humility in the face of surrounding circumstances. Instead of aesthetic values or God, the non-representationalist would frame moral perception as a sense of awe produced from surmising the complexity of materiality. Therefore, an individual in and of itself does not and cannot know the meaning of morality or social justice. Only by trusting and refining a discerning perception can an individual handle ever more complex relational processes in a way that eventually leads to acts we call moral and just. This non-representational approach that foregrounds situational particularities rather than pre-formulated moral rules supports previous attempts to formulate an open-ended, unanthropomorphized and unanthropocentricized outlook on moral agency within ethical theory (e.g., Birch, 1993; Blum, 1991; Cheney & Weston, 1999) by rendering their claims more substantial and empirically well founded.

The stance on social justice

In this stance on ethics, justice is conceived as a moral virtue, a character trait that enables a person to perceive as poignant the situational particularities that express (in)justice. This is the classical Aristotelian (Nichomachean) take on justice. Social justice, then, appears as a public good that regulates relations between people in a society in which conflicts may appear to result from a lack of resources and good intentions, or a disregard of the interests of others (Aristotle, 1993, pp. 1129a 26–1130a 26). This notion of genuine justice is pluralist to a certain degree, but it largely presumes a regard for the interests of others of a *kind*, in which proportionality is central. Nichomachean ethics takes into account the perspective of many parties so that proportionality of relations between those parties can be achieved (Aristotle, 1993, pp. 1129a 26–1129b 10).

The idea of proportionality as necessary to a formulation of justice pre-dates Aristotle (think of 'an eye for an eye') and infuses contemporary theories of justice. For example, the expression 'to get what is one's due' reflects how fundamental proportionality is to conceptions of justice. John Rawls' (1971) liberal theory of justice employs the maximin principle to offer an intricate algorithm of fair proportionality. Karl Marx's (2001[1875]) justice principle of contributing according to ability and receiving according to need is rooted in the idea of proportionality. And Iris Marion Young's (1990) call for justice as a lack of structural oppression for all social groups also maintains a commitment to proportionality. In all of these instances, creating harmonious and just relations in a society is seen to involve proportionality. Any notion of mobility justice, therefore, will benefit from a consideration of the relationship between proportionality and justice.

I use the concept of 'turbulence' to frame proportionality within mobility justice. Turbulence has been used within mobilities studies to theorize a mesophysical outlook on matter, the human condition and mobility. Cresswell and Martin (2012) suggest that turbulence is a kind of mobility that supports an (epistem-)ontology of uncertainty, chance and particular and unique instances. Even when the existential condition of mobility is situated in a world of turbulence, I suggest there is a place for justice. One of the main obstacles in understanding the parameters of justice in a turbulent world within moral philosophy is hegemonic representationalist interpretations of intentional action and the nature of autonomy. To bypass this problem in building a framework of mobility justice in a turbulent world, I outline the representationalist approach to intentional action, and I then provide a non-representationalist alternative.

The representationalist approach to intentional action

Mobility, of course, involves moving from point A to point B, but it also involves meaning-making, embodiment and experiences (Cresswell, 2010). For such terms to make sense, mobility must be distinguished from movements that are simply part of a line of causality, like the movement of asteroids and footballs. Movements that are not initiated by physical causes are the result of intention. Causality may be the reason you drop the china plate, but intention is the reason you pick up the pieces and glue them back together. Thus, autonomy, a will to alter the world at hand, is closely associated with intention. Intention and autonomy also are understood as criteria for moral acts; predestined or random acts cannot, by definition, be moral. There is no genuine social justice if causality is all there is.

Conventional, and typically anthropocentric, moral philosophy is allied with a representationalist outlook on autonomy and intention. According to the representationalist interpretation of the standard model of intentional action, such action is closely associated with practical reason (Stoutland, 2011). Intention is understood as a judgment attached to a possible action that is deemed desirable or good by particular terms (Davidson, 2002, p. 101). Judgments that are intentions are typically associated with wanting to have certain wants (e.g., wanting to like sweets less). If such second-order volition is informed by practical reason, then we have an intentional action that, furthermore, is autonomous (to which moral responsibility is attached). While the criteria for intentional action include references to reason, and only human beings are believed to have reason, personhood has only been granted to human beings in this representationalist account (Frankfurt, 1971), meaning that only human beings may be included in the scope of justice. To avoid conceptually restricting intention to anthropocentric notions, I turn to a non-representationalist account of intention, which does not rely on a notion of practical reason.

A non-representationalist approach to intentional action

There is no reason to challenge the standard model of intentional action altogether. Human beings' intentional acts may very well be motivated by desires that are informed by beliefs, some of which are grounded in a better way than others. Rather than denying this, non-representationalist theory provides other accounts of the meanings of 'beliefs', 'desires' and 'better', which do not limit intentional action to human agents.

Silberstein and Chemero, for example, claim that "intention must not be merely causally prior to the action but must somehow correspond to the intentional structuring of action, *without being something over and above the action*" (2011, p. 14; emphasis added). While this non-representationalist understanding of intention mainly harmonizes with the standard model, it departs from it by conceiving intention as directly associated with material constellations. Cary Wolfe's reading of the linguistics of Jacques Derrida and Luhmann's systems theory deconstructs intention by rejecting the idea of logical orders and, consequently, the idea that intention is associated with second-order volitions. Instead, he views intentional actions as a selection of all kinds of actions (Wolfe, 2010, pp. 31, 41–43). There is no set hierarchy of action that positions certain wants as inherently more base or unreflective than others. Wolfe (2010, pp. 45–47) locates intention in the tension between the being that wants and what is wanted. Consciousness (beliefs, desires) does not harbor intention as such, but language, as a mode of successful communication, does (Wolfe, 2010, pp. 45–47, 56). By developing language, things are, literally, created (Wolfe, 2010, pp. 60–61).

Silberstein and Chemero's (2011) non-representationalist account of intention is more concrete. They understand cognition as constituted by environmental components, as well as aspects of the cognizing individual body. Preferences and preferred objects are created through the intra-action of an information-processing body and its environment. For example, the pretense of eating broccoli not only co-creates the notion and manner of speaking of broccoli as edible, but also makes something like broccoli exist. Just as the fleshy leaves of the original *Brassica oleraca* invited the pretense to breed cultivars, the allegedly cancer-fighting biochemistry of broccoli invites scientists to breed 'superbroccoli'. Intention is expressed through the relational aspect of this (quite long) process. Silberstein and Chemero (2011) mainly refer to the micro-scale adaptations between animal and niche, by way of sensorimotor abilities and nervous system. Wolfe uses Luhmann to emphasize the evolutionary scale of this process, which is understood to include more than genetic processes. But Silberstein and Chemero make a similar point. The preference to eat (something like) broccoli may be understood as the result of agricultural processing as much as an evolutionary co-creation of cognizing eater and environment. The notion of broccoli as edible is more fleeting than its actual edibility, but both notion and fact are co-created within the same dynamic system that includes agent and environment.

Silberstein and Chemero (2011) suggest that intentions are "order parameters" that make a cognizing system act within certain constraints. The agency of the system is channeled by these constraints towards certain acts. For a mouse in a labyrinth, the walls and corridors, the smell of food at the exit and prior memories are among the order parameters that channel the mouse towards finding the way out. For more complicated acts, say parenting a child, social expectations, the needs of the child and parent, the extent and kind of communicative skills, the socio-economy and more shape each caring situation towards certain acts. The agency of the mouse or parent gains the honorary title of 'intentional' if it is restricted by the organism's attempt to make sense of its environment, to find meaning or to acquire a sense of value (Silberstein & Chemero, 2011).

If a severely stressed mouse enters the labyrinth, then the lure of food might not be strong enough to 'order' the acts of the mouse, so it may remain passive or display stereotypies. The mouse would not be in a state of being that enables it to even attempt to find meaning in the situation; its behavior would not be intentional. Analogously, parenting may involve losing one's temper, at which point the usual attempts to find meaning in being a parent, to make sense of the child's attempt to communicate or to value the child are temporarily disrupted. Behavior on such occasions may be channeled by situational particularities (parameters), but it is not intentional.

There are, then, 'better' ways to behave. Performances imbued with successful modes of communication with meaningful consequences are better than others. This account of intention may offer an ethical foundation for non-representationalists, but it also renders unclear whether morality is merely a structural quality of the system or if the individual is implicated. In other words, autonomy is at stake.

A non-representationalist account of autonomy

For the non-representationalist approach to intention to be ethically convincing, it needs to offer an account of why we would believe that individuals, at least occasionally, are responsible for their acts. Otherwise, individual and moral responsibility is in question and, with it, the very idea of justice. Immanuel Kant understands autonomy as inseparable from practical reason. He acknowledges the ever-changing nature of matter, but uses the idea of practical reason to explain how people gain stable insights into what is 'right' and transcend the variability of their bodies, social relations and environment (Kant, 1914[1788], pp. 27, 31–32). This approach to autonomy structures the representationalist interpretation of the standard model of intentional action.

However, non-representational theorists eschew the concept of practical reason (Wolfe, 2010, p. xxv) in favor of the view that the sensation of intentionality is a human illusion, and acts are founded in precognitive relational processes (Thrift, 2008, pp. 6–9, 166). To replace the Kantian notion of autonomy, non-representationalists provide an account of beings' capacity to choose well as part of becoming in the world. For example, Niklas Luhmann's (1988) principle of openness from closure is central to Wolfe's (2010, pp. xxi–xxv) outlook on autonomy. The autopoiesis of an organismal system closes such a system to itself, including the content of speech and cognition. If this closure were all there was, then the concept of autonomy would be no more than a majestic illusion veiling subjective prattle and predestined or random acts. It is assumed, however, that energy escapes the self-referential bubble.

Any self-sustaining system is dependent on energy input from its environment. The environment that harbors and provides energy to the organism offers overwhelming meaningfulness and material resistance, which must be handled by the organism, at the very least, to the extent of survival. This state of dependence encumbers the organism with adaptations and memories that create domains of information about the environment, which it shares with other organisms. Such domains are the foundation of communication (Wolfe, 2010, p. xxiv).

Individual organisms embody memories and adaptations of former intra-actions with certain aspects of the environment, and express them by communicative habits. As more environmental aspects are incorporated into the embodied self, the more complicated become the acts performed by the organism in that environment (Wolfe, 2010, p. xxi). Therefore, the appearance of the organism itself, seemingly independent from the environment, is a consequence of its dependence on, growth with and continuous intra-action with that environment. Autonomy accumulates with the complexity of the organism's intra-actions with its environment. Possibilities for autonomous action are temporary and conditional.

Silberstein and Chemero (2011) suggest that the complexity and non-causality necessary for achieving autonomy are reflected through the plasticity and robustness of the organism. Phenotypical plasticity means that genetically identical animals may develop disparate phenotypes due to differences in environment. Robustness entails certain traits of an organism persisting in spite of environmental and genetic variations. Both plasticity and robustness point toward the organism constituting a system in and of itself. Its resilience against certain genetic and environmental changes within certain limits admits a degree of individuality. This individuality grants the possibility of autonomy, which here refers to the ability to deploy various processes in order to uphold suitable relations in particular environmental circumstances. Such relations may be experienced as meaning and value (Silberstein & Chemero, 2011).

Just as the possibility for intentional action may be disrupted by changes in situational particularities, so can the autonomy of an organism. Responsibility for certain acts may be questionable in the non-representational view. But the moral agent is still obliged to intra-act and is responsible for intra-acting in its environment and for developing an ability to be autonomous, perceptive and considerate whenever such opportunities present themselves (Thrift, 2008, pp. 285–287). Such an attitude gains meaning from the natureculture that is harboring the organism, which becomes proficient by accumulating embodied memories and adaptations of intra-actions with its environment. Having such an attitude is right because it readies the moral agent's capacity to make sense of the overwhelming meaningfulness of the environment, to discern values and to orient its actions according to such values.

Turbulent moral agency

The turbulence discourse questions the dichotomous relation between order and disorder, and between stasis and mobility. The robustness of an organism, for example, is upheld by myriads of chemical reactions in ever-flowing bodily fluids and intra-actions with habitats. Phenotypical plasticity rests on a capacity for morphological disorder, but becomes ordered through the organism's continuous attempts to handle environmental stress. Order and disorder are braided in ever more complicated patterns until the autonomous body emerges in spite of being networked into environmental and genetical parameters. This process entails the labor of preparing the birth of morality thanks to the 'irreducible complexity in the order of events' in a world of turbulent matter (Webb, 2000, p. xii).

The braiding of plasticity and robustness is analogous to Michel Serres' (2000, p. 27) example of turbulence; the flow of a river is ordered only because of the overwhelming force and complexity of many disordered flows. There are times when the moral vortex gets to swirl and certain organisms become autonomous moral agents; those are times when situational particularities allow for moral agency to be exercised. At other times, material constellations forbid the very same organisms from exercising their moral agency. The turbulent nature of opportunities for autonomy means that whoever gains moral agency must be considered unknown until such agency actually appears.

Concluding discussion: turbulent justice in a more-than-human world

Recollect, then, that mobility is existential and relational for any mobile agent, no matter their species, and it may be conceptualized as a capability intrinsic to dignity that is owed such agents. Justice includes a notion of proportionality and relies on the virtuous ability to perceive proportional and disproportional distribution of those goods we owe each other. A straightforward conclusion, of course, is that mobility justice would prescribe the proportionate distribution of the capability of mobility among the objects of justice. From the analysis above, however, mobility has a more complicated relation to justice than merely being a good to be distributed. As just acts presume moral agency, as the development of a body into an autonomous organism relies on adapting to the environment and as mobility expedites the ability to manage the complexity of the environment, mobility facilitates the development of the virtue of justice. The proportionate distribution of the capability of mobility, therefore, is necessary to affirm the turbulent nature of those beings that may gain from having such a capability and to allow them to develop into agents and even, perhaps, moral agents. The embodiment of moral agency and the capability of mobility are mutually supportive.

Such embodiment of moral agency is not necessarily limited to one species. Indeed, most primates refuse to cooperate when the reward is unequal to the acquired reward of a visible conspecific (Price & Brosnan, 2012). Similarly, dogs refuse or hesitate to cooperate under conditions of unequal distribution (Horowitz, 2012; Range, Leitner & Virányi, 2012). Ravens seem to uphold rules of food ownership through third-party ravens (not affected by the theft) that punish rule-breaking individuals (Heinrich, 2006, pp. 269–279). Rats invest extra work in order to share a food source with conspecifics (Bartal, Decety & Mason, 2011). The tendency towards this type of behavior varies not only between species, but also within species (Price & Brosnan, 2012). To what extent terms like 'fairness' or 'justice' can be used to describe these behaviors is still debated (Brosnan & de Waal, 2012; Christen & Glock, 2012; Yamamoto & Takimoto, 2012). The point, however, is that the representationalist habit of logically restricting intentional action and autonomy to human beings is presumptuous and may belie empirical evidence. The question of who has moral agency is a matter of empirics, not logics, and non-representationalism offers apt openness to the concepts involved.

This ethological research demonstrates that a being apparently capable of intentional action and of perceiving proportionate and disproportionate distributive patterns may only concern herself when she is at a distributive disadvantage – apparently lacking altruism. Such behaviors are commonly observed, for example, among Capuchin monkeys (Yamamoto & Takimoto, 2012). Whether these observations are due to methodological limitations or actual behavioral limitations is unclear. Regardless, the adaptations made while struggling with the complexity of the environment must create some degree of altruism in order for recognizable social justice to appear.

The attention-oriented ethics outlined in the introduction of the chapter offer an understanding of how a non-representationalist approach to intention and autonomy suggests the possibility of developing altruistic behaviors. Weil and Murdoch suggest that moral insights rest on percepts of transcendental values. Their approach to moral agency as fully dependent on the agent's protruding relations is similar to the non-representational outlook of moral agency. Instead of transcendental values, non-representational theory foregrounds adaptations of communicative habits or language. Instead of humility, it is more apt to speak of a willingness to communicate. Herein lies the potential development of altruism.

Communication only makes sense if there is an entity to communicate with. Such an entity must be perceived as being a communicable significant other. While communicative habits are the source of moral agency, the treating of a communicable entity as *someone*, rather than *something*, would be a necessary consequence of developing agency. As communication becomes more complicated, "ideal role-taking" (Habermas, 1995) becomes more important. The agent's developing relation to the environment, including communicable entities, may reach a point where memories and adaptations have shaped communicative habits to the extent that the agent embodies some aspects of the reciprocity of another communicable someone and its environment. Agents' self-referential loops are linked by their common dependence on the environment, and they close in on each other's lifeworlds by perceiving the same world. The relational process between self-sustaining beings and the environment they are networked into may force on them the accumulation of communicative habits so complicated that role-taking becomes both possible and necessary. Many of the animal behaviors mentioned previously, like food-sharing, suggest that some degree of role-taking is common in species other than our own. From this insight on role-taking, we understand that those I communicate with share a somewhat similar situation in this world, and because we share this world commonly, they should be treated as I am. Proportionality in the division of labor, resources, punishment and merit may be expected within such communities.

Such an account of the development and unexpected appearance of intention and autonomy, which eventually conjoin with role-taking, has an important consequence for the scope of justice. It would be self-defeating to say that justice is owed only to those who have a sense of justice themselves. Rawls (1971, pp. 505, 512) suggests such a rationale based on the idea that justice is premised on mutual advantage. In contrast, I argue that once the virtue of justice is acquired, and at those flickering moments when it can be integrated into behavior, behavior necessarily adapts so that justice is done towards those that have something to gain from being treated justly. The scope of justice, then, would expand to include, at least, more animals than human beings.

What may be gained from being treated justly is unpredictable, so it is non-sense to include or exclude someone now based on the prospect that something may or may not happen in the future. Opportunities for mobility, however, should be granted proportionately and to an extent that respects the dignity of mobile agents. Some agents may then develop a sense of justice, but which ones cannot be known beforehand. Considering the development of moral agency, actual mobility has an instrumental value while it supports the development of social justice. Considering the scope of justice, however, the capability of mobility has an inherent value as an intrinsic part of the dignity of mobile agents. The reason why someone should be an object of justice is their dignity as an agent. The problem of marginal cases, that Nussbaum (2006) identifies in Rawls' theory – that the poor, disabled and nonhuman are excluded from the scope of justice – is avoided by this rationale. Even coma patients are included as their ability to move again must be considered unpredictable. By virtue of the minimal chance that they regain consciousness, they are owed respect at least to the extent of being cared for. What matters for inclusion into the scope of justice is the unpredictably mobile nature of a being.

This insight provides yet another argument against the absurdly contradictory and surprisingly persistent notion that moral agency and a sense of justice justify the oppression of those less capable. Exactly who should be included in the scope of justice, however, cannot be logically concluded, because turbulent agency is an empirical matter. Nevertheless, I believe it is vitally important not to presume that sentience is a prerequisite for turbulent agency. Hopefully, with the rise of a non-representational ethics, we will also see the demise of sentientism, which dominates animal ethics. Animals are worthy of respect by virtue of their (potentially mobile) existence, not because they can feel happy.

References

Adey, P. (2017). *Mobility*. New York: Routledge.

Aristotle. (1993). *Den nichomachiska etiken*. Gothemburg: Daidalos.

Bartal, I., Decety, J. & Mason, P. (2011). 'Empathy and pro-social behavior in rats'. *Science*, 334(6061), 1427–1430.

Bergmann, S. & Sager, T. (2008). 'Introduction'. In S. Bergmann & T. Sager (Eds.), *The ethics of mobility: Rethinking place, exclusion, freedom and environment* (pp. 1–9). Abingdon, UK: Taylor & Francis.

Birch, T. (1993). 'Moral considerability and universal consideration'. *Environmental Ethics*, 15(4), 313–332.

Blum, L. (1991). 'Moral perception and particularity'. *Ethics*, 101(4), 701–725.

Brosnan, S. & de Waal, F. B. M. (2012). 'Fairness in animals: Where to from here?' *Social Justice Research*, 25(3), 336–351.

Cheney, J. & Weston, A. (1999). 'Environmental ethics as environmental etiquette'. *Environmental Ethics*, 21(2), 115–134.

Christen, M. & Glock, H. (2012). 'The (limited) space for justice in social animals'. *Social Justice Research*, 25(3), 298–326.

Cresswell, T. (2010). 'Towards a politics of mobility'. *Environment & Planning D: Society & Space*, 28(1), 17–31.

Cresswell, T. & Martin, C. (2012). 'On turbulence: Entanglements of disorder and order on a Devon beach'. *Tijdschrift voor Economische en Sociale Geografie*, 103(5), 516–529.

Davidson, D. (2002). *Essays on actions and events*. New York: Oxford University Press.

Donovan, J. (1996). 'Attention to suffering'. *Journal of Social Philosophy*, 27(1), 81–102.

Frankfurt, H. (1971). 'Freedom of will and the concept of a person'. *Journal of Philosophy*, 68(1), 5–20.

Habermas, J. (1995). 'Reconciliation through the public use of reason: Remarks on John Rawls's political liberalism'. *The Journal of Philosophy*, 92(3), 109–131.

Heinrich, B. (2006). *Mind of the raven: Investigations and adventures with wolf-birds*. New York: HarperCollins Publishers.

Horowitz, A. (2012). 'Fair is fine, but more is better: Limits to inequality aversion in the domestic dog'. *Social Justice Research*, 25(2), 195–212.

Hume, D. (1998[1751]). *An enquiry concerning the principles of morals*. Oxford: Clarendon Press.

Kant, I. (1914[1788]). *Immanuel Kants werke: Bd 5, kritik der praktischen vernunft, kritik der urteilskraft*. Berlin: Cassirer.

Karlsson, F. (2012). 'Critical anthropomorphism and animal ethics'. *Journal of Agricultural & Environmental Ethics*, 25(5), 707–720.

Kronlid, D. (2016). 'Mobility as capability'. In T. Cresswell & T. P. Uteng (Eds.), *Gendered mobilities: Towards a holistic understanding* (pp. 15–34). New York: Routledge.

Luhmann, N. (1988). 'Closure and openness: On reality in the world of law'. In G. Teubner (Ed.), *Autopoietic law: A new approach to law and society* (pp. 335–348). Berlin: Walter de Gruyter.

Martens, K. (2012). 'Justice in transport as justice in access: Applying Walzer's "spheres of justice" to the transport sector'. *Transportation*, 39(6), 1035–1053.

Marx, K. (2001[1875]). *Critique of the Gotha program*. London: Electric Book Company.

McNay, L. (2004). 'Agency and experience: Gender as lived relation'. *The Sociological Review*, 52(2), 175–190.

Merriman, P. (2012). *Mobility, space and culture*. New York: Routledge.

Murdoch, I. (2001). *The sovereignty of good*. London: Routledge.

Noddings, N. (2003). *Caring: A feminine approach to ethics and moral education*. Berkeley, CA: California University Press.

Nussbaum, M. (2006). *Frontiers of justice: Disability, nationality, species membership*. Cambridge, MA: Harvard University Press.

Nussbaum, M. (2003). *Upheavals of thought: The intelligence of emotions*. Cambridge: Cambridge University Press.

Price, S. & Brosnan, S. (2012). 'To each according to his need? Variability in the responses to inequity in non-human primates'. *Social Justice Research*, 25(2), 140–169.

Range, F., Leitner, K. & Virányi, Z. (2012). 'The influence of the relationship and motivation on inequity aversion in dogs'. *Social Justice Research*, 25(2), 170–194.

Rawls, J. (1971). *A theory of justice*. Cambridge, MA: Harvard University Press.

Sager, A. (2018). *Toward a cosmopolitan ethics of mobility: The migrant's-eye view of the world*. Cham, CH: Palgrave Macmillan.

Sen, A. (2001). *Development as freedom*. Oxford: Oxford University Press.

Serres, M. (2000). *The birth of physics*. Manchester, UK: Clinamen Press.

Silberstein, M. & Chemero, A. (2011). 'Dynamics, agency and intentional action'. *Humana.Mente*, 15, 1–19.

Stoutland, F. (2011). 'Summary of Anscombe's intention'. In A. Ford (Ed.), *Essays on Anscombe's intention* (pp. 23–32). Cambridge, MA: Harvard University Press.

Thrift, N. (2008). *Non-representational theory: Space, politics, affect*. New York: Routledge.

Webb, D. (2000). 'Introduction'. In M. Serres (Ed.), *The birth of physics* (pp. vii–xx). Manchester, UK: Clinamen.

Weil, S. (1972). *Gravity and grace*. London: Routledge and Kegan Paul.

Wolfe, C. (2010). *What is posthumanism?*Minneapolis, MN: University of Minnesota Press.

Yamamoto, S. & Takimoto, A. (2012). 'Empathy and fairness: Psychological mechanisms for eliciting and maintaining prosociality and cooperation in primates'. *Social Justice Research*, 25(3), 233–255.

Young, I. M. (1990). *Justice and the politics of difference*. Princeton, NJ: Princeton University Press.

16 Tick movements
Patterning multispecies vulnerabilities

Jacob Bull

Introducing TBEv: 'going viral'

Mobility justice is always more-than-human in that it is rooted in ecological systems as well as being constituted with and through nonhuman agents (Sheller, 2016). That said, recognizing the more-than-human poses problems for concepts of inequality and justice. The humanist frameworks on which such concepts are based often fail to adequately attend to questions animal bodies pose by their presence. Further, and as Derrida famously reminds us, there is considerable diversity within the category of Animal (2008). In this chapter I address questions raised by this diversity by examining how vulnerabilities of humans and animals emerge through dynamic interactions between species that contribute to the production of tick-borne encephalitis (TBE) vulnerabilities. In so doing, I discuss relationships between people and animals that are always unequal (Haraway, 2008). Given this endemic inequality, the frame of 'justice' for this essay is about looking for ways of living (and dying) 'well' together (Haraway, 2008). One aspect of this is to avoid making individuals and groups what Haraway calls "killable" (i.e., easily, unreflexively, thoughtlessly killed). Another is to analyze how human–animal interaction contributes to the production of vulnerabilities for different groups of animals as well as humans. The third aspect is to recognize that animals are more than 'something worked on' by humans but also 'do work' in the production of vulnerabilities.

In this chapter I examine issues of living and dying together in the everyday context of Northern Europe. I focus on TBE vulnerability by asking how people and other animals are affected by TBE virus (TBEv) mobility. I engage texts that tell stories about the movements of animals and people, the processes by which TBEv passes between different bodies and the emotions and affects that circulate between and adhere to bodies. This storytelling matters because how TBEv is mobilized – materially, affectively and representationally – makes people and animals differently vulnerable. Vulnerabilities are produced through movements that are pathologized, rendering particular human and nonhuman beings 'at risk' and subject to different forms of governance. Ticks in these stories are always killable, but as I show, other bodies also become 'at risk'. The availability of TBE vaccines offers a way of living with ticks (and TBE), but the circulation of affects in vaccination advertising contributes to the pathologizing of particular bodies and movements. My focus on the nexus of bodily movement, viruses, affect and justice is informed by the "mobilities turn" (Urry, 2007) and attends to animal movements not as 'black boxes' but as productive and constitutive of 'things' and 'relationships' (Sheller & Urry, 2006), in this case TBE risks and vulnerabilities.

TBE is an infection caused by the TBE virus. According to the World Health Organization (WHO, 2011), TBE is "a serious acute central nervous system infection which may result in death or long-term neurological sequelae in 35–58% of patients. The fatality rate associated with clinical infection is 0.5–20%". There is no cure. With between 5,000 and 10,000 confirmed cases each year (Fritz et al., 2012), this infection is a widespread issue in Northern, Central and Eastern Europe (Figure 16.1). Indeed, it is one of the most significant viral-pathogenic threats to humans (Barret, Schober-Bendixen & Ehrlich, 2003). A widespread immunization program has been established in Austria, Germany, Sweden and other countries where it is considered endemic (Kunz & Heinz, 2003).

TBE is a growing phenomenon in Sweden. The first clinical case was identified in 1954 (Lundkvist, Wallensten, Vene & Hjertqvist, 2011). Incidence is still relatively low – around 257 cases per year – and the rate of increase is also slow (the mean between 2008 and 2012 was 236).[1] In Sweden, there are two to three clinical infections per 100,000 population, which is considerably more than in Germany or Finland (about 0.6/ 100,000), but less than in Estonia, Latvia and Lithuania (about 12/100,000) (ECDC, 2012). One response to TBEv presence in Sweden has been a growth in the vaccination program. Medical care is state supported in Sweden, but TBEv vaccinations are not state funded. A single dose costs 300–350 SEK (US$35/€30). Three doses are required for protection, followed by boosters every three to five years. Despite the cost, the annual vaccination rate over the last 20 years has risen from 50,000 to 600,000 doses (Slunge, 2015). The incidence is slowly growing, but the visual depiction of ticks and TBEv risks is such that TBEv is increasingly becoming part of the everyday landscape in Sweden. This growing 'everydayness' of TBEv risk is occurring despite the virus' presence in Sweden for more than 60 years and an effective vaccination that has been available since the 1980s. Why, therefore, has TBEv risk, to use Lavau's phrase (as cited in Sheller, 2016), 'gone viral' now?

To answer this question, I present TBEv risk as a natural-cultural phenomenon with specific geographical dynamics. Thereafter, I emphasize how movements of animals and people contribute to the emergence of TBEv as risky, producing particular bodies as 'at risk' and resulting in changing governance. This is in line with recent debates about the geographies of infection

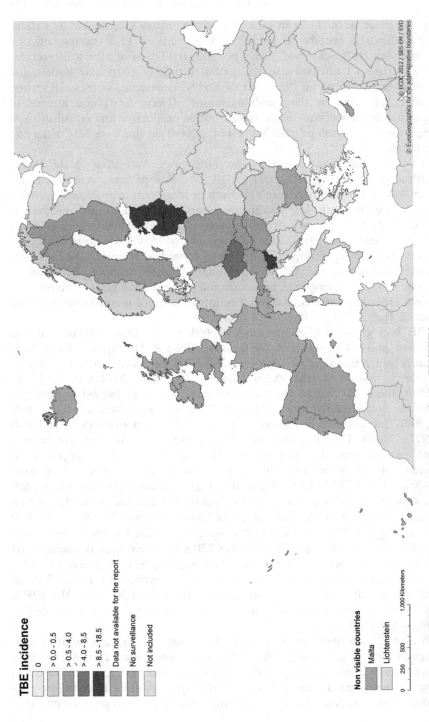

TBE incidence

- 0
- > 0.0 – 0.5
- > 0.5 – 4.0
- > 4.0 – 8.5
- > 8.5 – 18.5
- Data not available for the report
- No surveillance
- Not included

Non visible countries
- Malta
- Lichtenstein

0 250 500 1,000 Kilometers

© ECDC 2012 / SRS-EM / EVD
© EuroGeographics for the administrative boundaries

Figure 16.1 TBE average incidence per 100,000 inhabitants in the EU/EFTA
Source: ECDC (2012).

(e.g., Hinchliffe, Allen, Lavau, Bingham & Carter, 2013; Lorimer, 2016, 2017) that criticize conceptual frameworks which imply a 'pure' interior and 'contaminated' exterior. In contrast, Hinchliffe et al. (2013) argue for more nuanced approaches that are sensitive to the cultural, biological and technological ways that infections emerge in situated contexts.

In a description of the emergence of disease in intensive chicken production, Hinchliffe, Bingham, Allen and Carter (2017) identify supply and production chains, combined with economies of scale, casual labor, just-in-time logics and profit margins as contributing factors. These factors affect the interaction between human and avian bodies, producing vulnerabilities to infection. Similarly, I view TBEv risk and vulnerabilities not as a natural or social object, but as "the outcome of a *continuous interplay* between animals, microbes, people and materials that intra-act as they circulate, producing shifting and more or less pathogenic landscapes" (Hinchliffe et al., 2017, p. 101, emphasis in original). Therefore, my analysis foregrounds the movements of humans and animals, which combine with social structures, technologies and affects to create 'pathogenic landscapes'.

I begin by positioning tick movements in relation to the mobilities, mobilities justice and animal studies literatures. The first substantive section reviews the socio-geographical context for the emergence of TBEv as a 'threat', while the second addresses the movements of animals (and humans) that go into the making of the complex 'naturecultures' that mobilize TBEv. These sections illustrate the interplay between microbes, animals and people and highlight the implications of different ways of knowing for the production of vulnerabilities and their governance. In the third section I use Ahmed's (2004) notion of "affective economies" to discuss how affect and emotion are circulated in visual depictions of ticks and advertisements for TBE vaccinations. This section emphasizes the affective register that shapes TBEv vulnerabilities and argues that, as affects adhere to their bodies, ticks carry much more than TBEv.

More-than-human movements and mobilities

How ticks move, how they are moved and how they move others is at the core of this chapter. By considering ticks as agents in and subject to movements, I contribute to a growing discussion in the mobilities literature about more-than-human movement and mobilities. Recent work has focused on how animals and the products of their bodies are circulated around the globe. The 'horsemeat scandal' in Europe (Cresswell, 2014) and the histories of milk production that resulted in the relocation of cattle (Nimmo, 2011) are prime examples of these more-than-human mobilities. Additional work shows how animals, as they move between different spaces, shift between categories: dogs move from pets to threats (Instone & Mee, 2011), and rats change from pets to vermin to laboratory test subjects (Birke, 2003). Furthermore, animals are part of everyday human navigation; just considering dogs, we have guide dogs, sheep dogs, tracker dogs, sniffer dogs, hunting dogs, guard dogs, stray dogs, sled dogs and pet dogs. Animals are active companions in the making and navigating of space and the mooring of people (see Weaver, 2013, for an excellent discussion of how human and dog bodies mark each other, thereby altering their shared mobilities).

Biosecurity protocols – at national and local levels, on farms and in laboratories – regulate who and what can move, where, when and how. They further illustrate the ways that human and nonhuman movements and mobilities are tied together. This is particularly well illustrated in Sheller's (2016) description of how the cholera outbreak in Haiti was tied to human movements as vulnerable water systems were connected to distant infected areas through a network of airplanes, boots and equipment of international aid agencies. Recognizing the more-than-human aspects of human mobility has resulted in a call for "a deeper ecologizing of the material resource bases of mobility" (Sheller, 2016, p. 16).

A complementary step to ecologizing human mobilities involves attending to the mobilities of animals, not only how they facilitate human mobilities and moorings, but also how they are themselves fixed or mobile. This move recognizes that animals have their own navigations that may conflict with, undermine or exceed exclusively human conceptualizations and practices of movement. Mavhunga (2016), for example, focuses on the ways that tsetse flies are the mobile system connecting bodies across spaces and social and species divides. The agency of tsetse flies is demonstrated by the ways – because of the pathogenic networks they mobilize – they "force human societies to respond in multiple ways" (Mavhunga, 2016, p. 76). I follow Mavhunga's view of the vector of disease as 'a body at work' and, as such, see its movements as productive in ways that the mobilities turn has emphasized for humans. As demonstrated in the next two sections, tick movements are productive in that they contribute to the emergence of vulnerabilities and shifting forms of governance. But ticks also 'do work' as they carry with them more than microbes. Affect and emotion are so tied up with tick movements that it is important to engage with the more-than-representational (Lorimer, 2005) aspects of mobility.

As numerous scholars have shown, affects and emotions are political (Thrift, 2008; Vannini, 2015); they are the basis for why we act and move. Tick movements are no exception. Indeed, ticks generate an 'excess' of emotion. They are constructed as the ultimate other to human life and are discussed in the registers of disgust, fear and horror (Bull, 2014). Ticks do pose risks to human health. However, thousands of people live with bee and wasp allergies, and stings account for one to three deaths per year in Sweden (Gülen & Björkander, 2016). Despite this, bees and wasps do not invoke such widespread negative affects. It is therefore fair to say that there is a politics to animal movements as particular movements are pathologized. Tick movements are framed as 'undesirable' and, therefore, can be included in what

Söderström, Randeria, Ruedin, D'Amato and Panese (2013) term "critical mobilities".

Söderström et al. (2013, p. vi) suggest that analyses of mobilities are critical if they engage with "problematic" mobilities, (methodologically) challenge the limits of mobility in understanding spatial relations or critically attend to how mobilities shape social structures. Tick movements are considered by people to be problematic. In addition, focusing on these movements is critical to address the implications movement has for our more-than-human social worlds. I focus on the production of vulnerabilities through movements that are pathologized in various ways, rendering particular human and nonhuman beings 'at risk'. This final point connects to the core theme of this volume: mobility justice. Whereas a critical approach to mobilities engages how mobilities shape social structures, mobility justice asks more pointed questions about who and what can and cannot move, and how mobility as a socially differentiated practice produces and is produced by power (Cook & Butz, 2016, p. 401).

In the following sections I show that people are not the only powerful actors in the production of mobility. Following Mavhunga's (2016, p. 77) work on tsete flies, I focus on the ways tick bodies 'do work' within the social, biological, technological systems that produce vulnerabilities. This work becomes particularly evident in the ways that animals move between different categories, become valued differently and are subject to different forms of (human) governance because of the circulations of TBEv. However, to label these shifting forms of governance 'unjust' hides the complexity of the relations involved in finding ways to 'live well' in ecological systems. Vaccination poses an alternative route, but as I demonstrate in the chapter's final section, vaccination as a mode of living with ticks and TBEv is also fraught with asymmetries of power. Within the vaccination framework, ticks also do work as actors in the circulation of affects; they are what Ahmed (2004) calls "sticky" objects. Their 'stickiness' causes affects to adhere to their bodies, once more pathologizing them. This entanglement of affects, representations and practices of movement is key to the production of more (or less) 'pathogenic' landscapes.

Pathogenic landscapes: social and cultural patterning of TBE vulnerability

TBEv vulnerability in Sweden has a specific sociocultural geography. Although TBE is a biological phenomenon resulting from an infection spread by ticks, local and global geographies also shape how infection risks are assessed and vulnerabilities made. The TBE incidence mapped in Figure 16.1 shows that the disease is endemic to the Baltic states and Austria, but of decreasing concern in Western Europe, although its incidence is growing and spreading. Maps like this are representational mechanisms used to comprehend, organize and codify particular relations, so they contribute to the process by which TBEv vulnerabilities are made. Sweden sits in the >0.5–4 cases per 100,000 population category; five cases is the level at which the WHO recommends vaccination programs.

As the map visualizes this differential threat across Europe, it homogenizes risk within national boundaries. At a finer scale, as illustrated in Figure 16.2, we see more precise patterning. In Sweden, for example, TBE risk is predominantly coastal and in the central band across the country stretching from Gothenburg in the west to Stockholm in the east, with large parts of the country unaffected by TBEv. Figure 16.3 shows an even finer distribution of the confirmed cases in 2014.[2] At this level, TBE incidence becomes even more local, focusing on Stockholm, Lake Märlaren and the surrounding districts. In Figure 16.2, TBEv incidence is presented at the county level, distributing risks over a wider area, 'hiding' local risk and creating seemingly less pathogenic landscapes. As such, as Beck (1989) argues for risks more generally, TBE risk is not exclusively natural or social. Instead, it can be understood as a hybrid outcome of various intersecting policies, technologies, economic factors and decision-making processes.

Although maps contribute to the making of TBEv risk, they do not coincide with the risks individuals face. Slunge (2015) recognizes that living in a high TBE incidence area does not automatically increase vulnerability. There are additional cultural practices that increase the likelihood of infection. Slunge addresses how individual interests contribute to risks, with the relatively blunt tool of 'time spent outside'. Other studies link higher risk for tickborne illness with dog walking and mushroom picking (Jaenson, Hjertqvist, Bergström & Lundkvist, 2012). Studies such as those by Slunge and Jaensson et al. identify TBEv risk as a spatially patterned phenomenon, with vulnerabilities influenced by culture and practice. However, to present risk solely as the product of medical governance or cultural practices of keeping pets or picking mushrooms would underrepresent the networks of human and animal movements that contribute to the mobilization of TBEv. TBEv risk is not only spatially patterned, but it is also a product of human and animal movement.

Humanimal movements: towards ecologies of infection

While the previous section identified the socio-geographical components in the emergence of TBEv risk, this section attends to animals as mobile agents 'doing work' in the production of TBEv vulnerabilities. Jakob von Uexküll's (2010, p. 7) description of the lifeworld of ticks focuses on a trio of stimuli – temperature, acid concentration and hairiness:

> The eyeless tick is directed to [her] watchtower by a general photosensity of her skin. The approaching prey is revealed to the blind and deaf highway woman by her sense of smell. The odor of butyric acid, that emanates from the skin glands of all mammals, acts on the tick as a

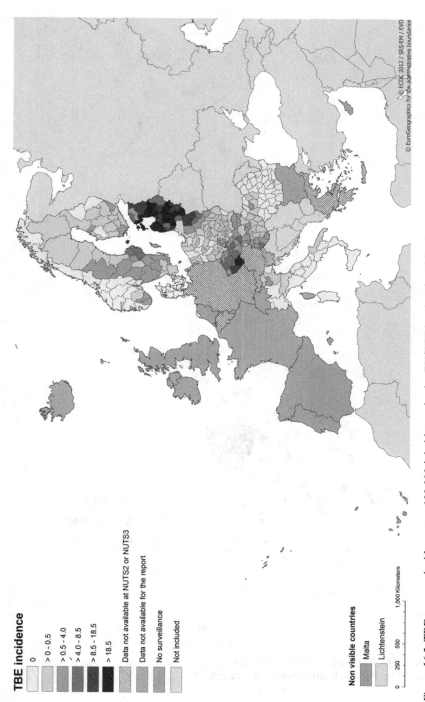

TBE incidence

- 0
- > 0 - 0.5
- > 0.5 - 4.0
- > 4.0 - 8.5
- > 8.5 - 18.5
- > 18.5
- Data not available at NUTS2 or NUTS3
- Data not available for the report
- No surveillance
- Not included

Non visible countries
- Malta
- Lichtenstein

0   250   500   1,000 Kilometers

© ECDC 2012 / SRS-EM / EVD

© EuroGeographics for the administrative boundaries

Figure 16.2 TBE average incidence per 100,000 inhabitants in the EU/EFTA subnational level

Source: ECDC (2012).

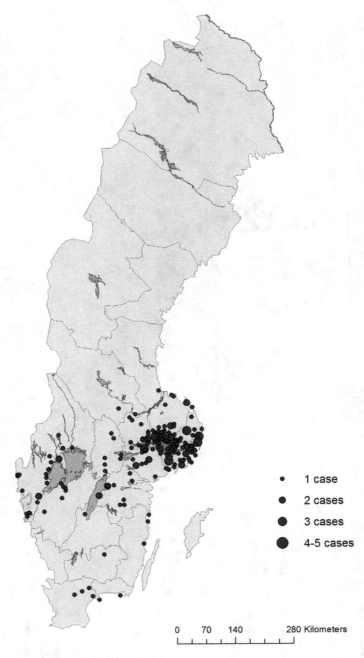

Figure 16.3 2014 TBE incidence, Sweden
Source: The Public Health Agency of Sweden.

signal to leave her watchtower and hurl herself downwards. If, in so doing, she lands on something warm – a fine sense of temperature betrays this to her – she has reached her prey, the warm-blooded creature. It only remains for her to find a hairless spot.

The geography of ticks described by von Uexküll is characterized by long periods of stillness with short, local movements: from bush to prey and across the body of the host. These movements are key steps in the mobilization of TBEv. However, I argue that a wider conceptualization of tick geographies is needed. For example, the movements of tick hosts expand the geographies of ticks exponentially, and studies have suggested that animal migrations are a key aspect in mobilizing TBEv in Sweden.

Coinciding with the relatively localized and coastal distribution of human TBE cases in Sweden (Figure 16.3), migrating birds have been proposed as a possible vector that carries the virus, the ticks or both across the Baltic Sea (Hasle, 2013). Waldenström et al. (2007) (Figure 16.4) show the migration routes of tree pipits, robins, redstarts and song thrushes, suggesting that their movements could contribute to the spread of TBEv. Roe deer have also been proposed as a factor in the mobilization of TBEv in Sweden as populations grow and ranges extend northward and into urban and peri-urban areas. Jaenson, Hjertqvist and Lundkvist (2012) explain that tick populations have grown in relation to deer populations as deer predator (e.g., fox and lynx) numbers decline due to an outbreak of scabies. These predators also prey on small rodents that are important hosts of TBEv.

The movements of birds, deer, lynx and small rodents are often presented as causal relationships. However, while linear causal links are proposed,

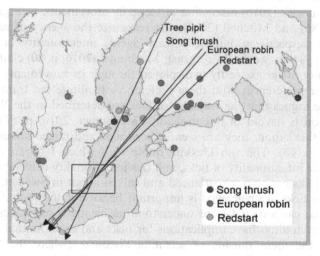

Figure 16.4 Mapping TBEv movements with birds
Source: Waldenström et al. (2007).

scientific accounts (e.g., Jaenson, Hjertqvist, Bergström & Lundkvist, 2012; Jaenson, Hjertqvist & Lundkvist, 2012) increasingly understand TBEv mobilities as resulting from complex interdependencies. In his review of Lyme disease risk, Ostfeld (2011) notes a pattern similar to that in stories of TBE outlined above, with causality variously attributed to deer, small rodents or the weather. He argues that the dynamics of Lyme disease are far more complicated than the linear stories in single host explanations, simplifications of tick life cycles and extrapolations of lab results that characterized early understandings of Lyme disease. Ostfeld (2011, p. 185) concludes that "infectious diseases are ecological systems" rather than causal relationships.

TBEv can be understood in a similar ecological form; many species intersect to provide multiple pathways for TBEv to move through spaces and between bodies. Birds or deer move ticks and TBEv across long distances. Deer and small rodents operate as reservoirs, and the tick's ability to go many years between blood meals (von Uexküll, 2010) means that there are infinite possibilities and permutations for TBEv to move with a variety of bodies in a variety of spaces and at a variety of speeds. Within this network, each species and individual animal have their own interrelated geographies and annual, seasonal and diurnal rhythms that are not defined by a single pathway. Therefore, no one organism is solely 'responsible' for increased risk (Ostfeld, 2011). Indeed, a multiplicity of actors – deer, mice, small mammals, birds, predators – become 'responsible' in different contexts and at different times, contributing to an ecology of disease. This ecological approach emphasizes biological complexity, but as Hinchliffe et al. (2017) argue, the emergence of disease also occurs in specific social and cultural contexts. There is a need to examine how ticks do work as they move within these social, ecological, technological networks.

In Mavhunga's (2016) discussion of tsetse fly movements, he argues against Callon (1986) and Mitchell (2002) who recognize the agency of scallops and mosquitoes respectively, but deny these agents intentionality (and thereby limit the ways they work). In contrast, Mavhunga (2016, p. 80) claims that the intentionality of the tsetse fly is displayed through its movements; it demonstrates clear preferences, and these preferences facilitate the transmission of disease. He argues that the ways tsetse flies are described in the Western scientific canon is devoid of intentionality (Mavhunga, 2016, p. 84), but that, outside of this canon, they are seen, experienced and talked about as having intentions (p. 78). The von Uexküll quote with which I began this section points to the intentionality of ticks, and this intentionality is recognized in the ways that ticks are seen, experienced and talked about in Sweden. Attending to the ways ticks are perceived is important because it shows that outside of the scientific cannon, ticks are understood as having 'preferences' and that these understandings have implications for ticks and other animals.

Shifting the focus outside of scientific discourse (where the 'ecological' perspective is more common), linear causal explanations continue to be circulated in media representations. For example, all of the following articles from national and local presses make connections between deer and/or bird populations and the growth or spread of ticks and TBEv (Anon SR, 2012; Anon SvD, 2012; Broström, 2016; Gad, 2011; Hermansson, 2017; Magnusson, 2016; Svärdkrona, 2012). There are presentations that take a more ecological perspective (e.g., Zuidervelt, 2013). However, the dominant way that ticks are talked about is as a 'lurking horror', and even the Zuidervelt article has the title "The experts' answers about the tick plague". Therefore, despite the science that points to an ecology of disease, public understandings of the intentionality of ticks combine with tick mobilities in ways that have implications for other animals.

The governance response to zoonotic disease usually results in the extension and normalization of human control over nonhuman life, as we see in the slaughter of animals after outbreaks of foot-and-mouth disease, avian flu and swine flu and regulation changes following BSE and scrapie outbreaks. A similar response to governing the animal bodies that work in the production of TBEv risk has arisen in Sweden. Some media articles (e.g., Svärdkrona, 2012) call for local elimination or reduction of the deer population as a mechanism for combatting TBEv. This is illustrative of how ticks as they move are bodies that work, recalibrating human-animal relationships such that animals also become differently vulnerable. In this instance, deer become disposable or "killable" (Haraway, 2008). In contrast, foxes and lynx become more valuable as they are understood to reduce 'problematic' deer and small rodent populations. A more-than-human politics of movement is evident as representations interact with practices of movement; tick mobilities become unwanted or 'critical' (Söderström et al., 2013), while others are valued; landscapes become less or more pathogenic accordingly.

Circulating affects: ticks moving humans

In the previous two sections I argued that many organisms interact with representational structures and cultural, medical and economic systems to produce TBEv risk, suggesting that the tick is a body that works, changing human-animal relationships as it moves. As ticks move, landscapes are rendered more or less pathogenic, and different organisms become killable or valuable. However, although many animals are implicated in the TBEv network, it is to ticks that TBEv most closely adheres. I use Ahmed's (2004) notion of "affective economies" to demonstrate how affect, animals and economies interact in the making of TBEv risk. In so doing, I address how affects, being emotionally moved, circulate and accumulate such that "emotions involve subjects and objects, but without residing positively within them" (Ahmed, 2004, p. 119).

The affective economies I discuss here recognize the work done by nonhuman bodies in the circulation of affect, making affect flow between humans and nonhumans. Slunge (2017) shows that people in Sweden are engaging in new practices (checking skin after walks and avoiding long grass and known tick-infested areas) because of their fear of ticks and tick-borne disease. Focusing on the marketing of TBEv vaccinations, I examine how tick movements become 'critical' as they are produced with and through 'negative' affect.

Drawing on Marx's notion that capital is accumulated through its circulation in commodity form, Ahmed (2004) attends to the ways that affect and emotion circulate and accrue. Because affect is not located *in* bodies but can be passed between, circulated and returned to bodies, emotions can "(re)produce or generate the effects that they do" (Ahmed, 2004, p. 124). Such circulation of affect is occurring in discourses of TBEv in Sweden. On the left in Figure 16.5 is a widely circulated image of a tick filling a traffic warning triangle. The image on the right is the headline from a tabloid newspaper: "Infected by the new horror tick". Ticks and the diseases they carry are clearly marked as something to fear. Ahmed highlights how bodies, in the circulation of affect, become "sticky", collecting emotive language as affect accumulates. Ticks have become sticky in TBEv discourses in Sweden, acquiring a red warning triangle and a link with infection and horror. Its lurking parasitic presence is represented in shopping center advertising (Figure 16.6) with the effect of reproducing a range of cultural anxieties around the parasitism of ticks.

The label 'parasite' is a powerful, instrumental mechanism of collective aggression to justify violence against human individuals and groups. Hollingsworth (2006) shows how insect metaphors have been prevalent in such violence. While ticks are arthropods not insects, as insect-like parasites they are one of the ultimate critters of *différance* (Sleigh, 2006). However, while humans have relatively narrow and dismissive cultural views of parasites, the latter "not only have far-reaching and important impacts on the earth's ecology, they have much broader effects on our daily lives than we realize" (Drisdelle, 2010, p. 2). Indeed, parasites are embraced as forms of biological control for agricultural crops and welcomed as companions in the management of human microbiomes (Lorimer, 2016). While some parasites are welcomed components of our multispecies worlds, ticks are constructed as the ultimate other to human ways of being and discursively connected to disease, horror and infection. They are, to use Ahmed's (2004) phrase, "sticky bodies".

This 'stickiness' is particularly evident in the naming of the disease. Unlike many bacterial or viral infections, TBEv is not named after a specific virus (e.g., influenza) or bacteria (e.g., *E. coli*); neither is it named after a particular person involved in initial diagnosis (e.g., CJD, Creutzfeldt-Jakob disease) or a place (e.g., Lyme disease) or social group (e.g., legionnaires' disease). Furthermore, TBEv does not affect only ticks, and so does not fit into the raft of diseases that are named after the species or group that are most affected by the disease (e.g., bovine TB). Instead of any of these naming systems, the TBE virus is named after the vector that passes it between different hosts – the tick. Ticks and the virus are firmly adhered.

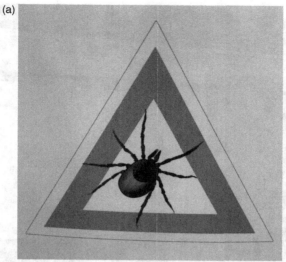

Figure 16.5 Tick warnings

The advertising for vaccination programs further demonstrates how emotions involve subjects and objects, but without residing positively within them (Ahmed, 2004) as they circulate negative affects. Vaccinportalen.se mobilizes the advertisements shown in Figures 16.5 and 16.6, while fasting.nu declares, "One tick bite is enough. Vaccinate the whole family against TBE".[3] In contrast to newspaper articles that present threats in terms of specific presences (e.g., deer or birds), these websites and vaccination programs frame the TBEv as a widespread issue. Playing on uncertainty, the circulation of affects

Figure 16.6 Adverts for TBEv vaccination in the Stockholm Underground (a) and a Gothenburg shopping center (b)

expands the threat beyond specific relationships, generating 'pathogenic landscapes'. While none of this is 'false' – infection sites/routes/times are uncertain and one bite is enough – ticks as sticky bodies 'do work' as affective agents to make landscapes 'more' pathogenic and vaccination relevant.

To return this description of the emergence of TBEv vulnerabilities to the social and cultural factors with which I began, the way that ticks do work as affective agents in the production of 'pathogenic landscapes' is particularly significant because the provision of TBEv vaccinations through market mechanisms is uneven. Current vaccination rates for TBEv are 15% for low-income and 50% for high-income groups, which Slunge (2015) attributes to vaccine pricing. I suggest that affect is operating in combination with ecologies and economies such that TBEv becomes an issue that warrants changes in behavior, increases in surveillance of our bodies and (as the marketing material for vaccinations implies) justifies the expenditure on privatized vaccination programs. It would be speculation to imply a causal link between affective economies surrounding ticks and private vaccination programs. However, I suggest that affect contributes to the naturecultures of risk that are produced in advertising for the vaccination program. The vulnerabilities that emerge are natural (in terms of a biological threat), social (in terms of bureaucratic decisions about national risk) and affective. In this way they are material, discursive and more-than-representational.

Conclusion: the mobilities and inequalities of TBEv vulnerabilities

As an answer to the question 'why has TBEv gone viral now?', I discussed the social geographies of TBE, the humanimal movements that constitute the ecologies of infection and the affects that are circulated and 'stick' to ticks. By discussing this tripartite structure of social, animal and affect, I have shown how human and animal movements contribute to the emergence of landscapes as 'pathogenic'. Starting from a mobilities perspective that views movements as productive rather than as 'black boxes', I have shown that humanimal movements produce consequences for people and animals. I have described how linear understandings of causality, linking particular animal movements with the spread of disease, makes animals differently vulnerable; deer or lynx (for example) move between categories such as valuable or 'killable'.

I have shown that mobility justice is 'more-than-human', both in terms of the ecological basis of the patterning of risk (i.e., uneven vulnerabilities are a consequence of social *and* natural processes) and the implications for both humans *and* animals stemming from the construction of risk. In my discussion of how landscapes are understood as pathogenic, I have highlighted the circulation and political power of affect. Using Ahmed's (2004) work, I discussed ticks as sticky objects that circulate and accumulate affect. This affect is circulated in a variety of media representations, including in the advertising of vaccination programs, which suggests that affective economies should be considered part of the emergence of TBEv risk and the 'pathogenic' landscapes of TBEv.

By focusing on the movements ticks make, shape and instigate, I have shown that animal bodies do work as they move. Mavhunga (2016, p. 88) argues for a wider conceptualization of 'vehicle' such that we might attend to the "many modes of auto-mobile (self-propelled) entities and what they carry". From the TBEv to affect, I have shown that ticks carry many things and do considerable work in the material, discursive and more-than-representational processes that produce 'pathogenic landscapes'. Searching for mobility justice for humans and nonhumans in these landscapes is difficult. Undoubtedly the emergence of TBEv vulnerabilities is uneven, has implications for the actors involved and affects how they can move. Rather than looking for justice *per se*, I set myself the task of looking at ways of living together in complex worlds of humans and animals. People are made vulnerable by the movements of humans and animals. Movements of ticks in combination with linear narratives, such as those in media representations, make other animals differently vulnerable or valuable. Vaccination offers possible routes for living together. However, here too ticks 'do work' as sticky subjects of an excess of emotion; they are active in the production of pathogenic landscapes. Ways of living and dying together are always mobile and asymmetrical.

Notes

1 Retrieved from the website of the Public Health Agency of Sweden. www.folkha lsomyndigheten.se/folkhalsorapportering-statistik/statistikdatabaser-och-visualiser ing/sjukdomsstatistik/tick-borne-encephalitis-tbe/?t=county.
2 Retrieved from the website of the Public Health Agency of Sweden. www.folkha lsomyndigheten.se/nyheter-och-press/nyhetsarkiv/2015/april/antalet-fall-av-tbe-m inskade-forra-aret/.
3 See www.vaccinportalen.se and https://fasting.nu/?gclid=EAIaIQobChMI2IC0op SO1wIVR-AYCh2eagKKEAAYAiAAEgJr4_D_BwE.

References

Ahmed, S. (2004). 'Affective economies'. *Social Text*, 22(2), 117–139.
Anon SR. (2012, June 3). 'Forskare: Skjut rådjur och minska fästingarna'. *Sveriges Radio*. Retrieved from https://sverigesradio.se/sida/artikel.aspx?programid=125&a rtikel=5135472.
Anon SvD. (2012, July 6). 'Botemedel mot TBE närmar sig'. *Svenska Dagbladet*. Retrieved from www.svd.se/botemedel-mot-tbe-narmar-sig.
Barret, N. P., Schober-Bendixen, S. & Ehrlich, H. J. (2003). 'History of TBE vaccines'. *Vaccine*, 21(1), 41–49.
Beck, U. (1989). *The risk society*. London: Sage.
Birke, L. (2003). 'Who – or what – are the rats (and mice) in the laboratory'. *Society & Animals*, 11(3), 207–224.

Broström, L. (2016, July 25). 'Färre rådjur i markerna ger färre fästingar med borrelia'. *Vetenskapsradio.* Retrieved from https://sverigesradio.se/sida/artikel.aspx?pro gramid=406&artikel=6480127.

Bull, J. (2014) 'Between ticks and people: Responding to nearbys and contentments'. *Emotion, Space and Society,* 12, 73–84.

Callon, M. (1986). 'Some elements of a sociology of translation: Domestication of the scallops and the fishermen of St. Brieuc Bay'. In J. Law (Ed.), *Power, action and belief* (pp. 196–223). London: Routledge.

Cook, N. & Butz, D. (2016). 'Mobility justice in the context of disaster'. *Mobilities,* 11 (3), 400–419.

Cresswell, T. (2014). 'Mobilities III: Moving on'. *Progress in Human Geography,* 38(5), 712–721.

Derrida, J. (2008). *The animal that therefore I am.* New York: Fordham University Press.

Drisdelle, R. (2010). *Parasites: Tales of humanity's most unwelcome guests.* Berkeley, CA: University of California Press.

ECDC. (2012). *Epidemiological situation of tick-borne encephalitis in the European Union and Europena Free Trade Association countries.* Stockholm: European Centre for Disease Prevention and Control.

Fritz, R., Orlinger, K., Hofmeister, Y., Janecki, K., Trawaeger, A., Perez-Burgos, L., Barret, P. & Kreil, T. (2012). 'Quantitative comparison of the cross-protection induced by tick-borne encephalitis virus vaccines based on European and Far Eastern virus subtypes'. *Vaccine,* 30(6), 1165–1169.

Gad, M. (2011, July 15). 'Fästingen sprider sig över landet – samtidigt ökar tbe'. *Aftonbladet.* Retrieved from www.aftonbladet.se/nyheter/article13329627.ab.

Gülen, T. & Björkander, J. (2016). 'Insektsgiftallergi – diagnostiken kan vara svår men bra behandling finns'. *Läkartidningen,* 113, 34–35.

Haraway, D. (2008). *When species meet.* Minneapolis, MN: University of Minnesota Press.

Hasle, G. (2013). 'Transport of ixodid ticks and tick-borne pathogens by migratory birds'. *Frontiers in Cellular and Infection Microbiology,* 3(48). Retrieved from http:// doi.org/10.3389/fcimb.2013.00048.

Hermansson, S. (2017, July 21). 'TBE-läkaren: Ingen större anledning till oro'. *SVT Nyheter Helsingborg.* Retrieved from www.svt.se/nyheter/lokalt/helsingborg/tbe-laka re-efter-forsta-fallet-i-nordvastra-skane-ingen-storre-anledning-till-oro.

Hinchliffe, S., Allen, J., Lavau, S., Bingham, N. & Carter, S. (2013). 'Biosecurity and the topologies of infected life: From borderlines to borderlands'. *Transactions of the Institute of British Geographers,* 38(4), 531–543.

Hinchliffe, S., Bingham, N., Allen, J. & Carter, S. (2017). *Pathological lives: Disease, space and biopolitics.* Chichester, UK: Wiley-Blackwell.

Hollingsworth, C. (2006). 'The force of the entomological Other: Insects as instruments of intolerant thought and oppressive action'. In E. Brown (Ed.), *Insect poetics* (pp. 262–281). Minneapolis, MN: University of Minnesota Press.

Instone, L. & Mee, K. (2011). 'Companion acts and companion species: Boundary transgressions and the place of dogs in urban space'. In J. Bull (Ed.), *Animal movements, moving animals: Essays on direction, velocity and agency in humanimal encounters* (pp. 229–250). Uppsala: Uppsala University Press.

Jaenson, T. G., Hjertqvist, M., Bergström, T. & Lundkvist, Å. (2012). 'Why is tick-borne encephalitis increasing? A review of the key factors causing the increasing incidence of human TBE in Sweden'. *Parasites & Vectors,* 5, 184–197.

Jaenson, T. G., Hjertqvist, M., & Lundkvist, Å. (2012). 'År 2011 toppar TBE-incidensen: Rådjursstammens variation i storlek och vädret är nyckelfaktorer'. *Läkartidningen*, 109(7), 343–346.

Kunz, C. & Heinz, F. X. (2003). 'Editorial: Tick-borne encephalitis'. *Vaccine*, 21(1), 1–2.

Lorimer, J. (2017). 'Microbiogeographies: The lively cartographies of *Homo microbis*'. In J. Bull, T. Holmberg & C. Åsberg (Eds.), *Animal places: Lively cartographies of human–animal relations* (pp. 185–206). Abingdon, UK: Routledge.

Lorimer, J. (2016). 'Gut buddies: Multispecies studies and the microbiome'. *Environmental Humanities*, 8(1), 57–76.

Lorimer, H. (2005). 'Cultural geography: The busyness of being "more-than-representational"'. *Progress in Human Geography*, 29(1), 83–94.

Lundkvist, A., Wallensten, A., Vene, S., & Hjertqvist, M. (2011). 'Tick-borne encephalitis increasing in Sweden'. *EuroSurveillance*, 16(39), pii.

Magnusson, A. (2016). 'Tänk på att de bär med sig 1 000 fästingar'. *Gefle Dagblad*. Retrieved from www.gd.se/artikel/opinion/insandare/tank-pa-att-de-bar-med-sig-1-000-fastingar

Mavhunga, C. (2016). 'Organic vehicles and passengers: The tsetse fly as transient analytical workspace'. *Transfers*, 6(2), 74–93.

Mitchell, T. (2002). *Rule of experts: Egypt, techno-politics, modernity*. Berkeley, CA: University of California Press.

Nimmo, R. (2011). 'Bovine mobilities and vital movements: Flows of milk, mediation and animal agency'. In J. Bull (Ed.), *Animal movements, moving animals: Essays on direction, velocity and agency in humanimal encounters* (pp. 57–74). Uppsala: Uppsala University Press.

Ostfeld, R. S. (2011). *Lyme disease: The ecology of a complex system*. New York: Oxford University Press.

Sheller, M. (2016). 'Uneven mobility futures: A Foucauldian approach'. *Mobilities*, 11 (1), 15–31.

Sheller, M. & Urry, J. (2006). 'The new mobilities paradigm'. *Environment & Planning A: Economy & Space*, 38(2), 207–226.

Sleigh, C. (2006). 'Inside out: The unsettling nature of insects'. In E. Brown (Ed.), *Insect poetics* (pp. 281–297). Minneapolis, MN: University of Minnesota Press.

Slunge, D. (2017). *Essays in environmental management and economics: Public health, risk and strategic environmental assessment*, Published doctoral thesis, University of Gothenburg, Gothenburg.

Slunge, D. (2015). 'The willingness to pay for vaccination against tick-borne encephalitis and implications for public health policy: Evidence from Sweden'. *PLoS ONE*, 10(12). Retrieved from https://doi.org/10.1371/journal.pone.0143875.

Söderström, O., Randeria, S., Ruedin, D., D'Amato, G. & Panese, F. (Eds.). (2013). *Critical mobilities*. Lausanne: EPFL Press.

Svärdkrona, Z. (2012, January 16). 'Skjut bort rådjuren – bra för folkhälsan'. *Aftonbladet*. Retrieved from www.aftonbladet.se/nyheter/article14219657.ab.

Thrift, N. (2008). *Non-representational theory: Space, politics, affect*. London: Routledge.

Urry, J. (2007). *Mobilities*. Cambridge: Polity Press.

Vannini, P. (2015). 'Non-representational research methodologies: An introduction'. In P. Vannini (Ed.), *Non-representational methodologies: Re-envisioning research* (pp. 1–18). New York: Routledge.

von Uexküll, J. (2010). *A foray into the worlds of animals and humans, with a theory of meaning*. Minneapolis, MN: University of Minnesota Press.

Waldenström, J., Lundkvist, Å., Falk, K. I., Garpmo, U., Bergström, S., Lindegren, G., ... & Olsen, B. (2007). 'Migrating birds and tickborne encephalitis virus'. *Emerging Infectious Diseases*, 13(8), 1215–1218.

Weaver, H. (2013). 'Becoming in kind: Race, class, gender and nation in cultures of dog rescue and dog fighting'. *American Quarterly*, 65(3), 689–709.

WHO (World Health Organization). (2011, August 9). 'Immunizations, vaccines and biologicals: Tick-borne encephalitis'. Retrieved from www.who.int/immunization/topics/tick_encephalitis/en/.

17 Redistributing surplus food

Interrogating the collision of waste and justice

Anna R. Davies

Introduction

Food *in toto* is highly mobile. Not only does it get moved around from sites of production to sites of processing, consumption and, ultimately, divestment and disposal, as a biological entity it is also characterized by molecular mobilities as it grows, matures and decomposes (Davies, 2012). This movement has led some scholars to characterize food as a hypermobile object that is "good to think mobilities with" (Gibson, 2007, p. 4). However, while the dynamic processes of relocation (moving food from farm to fork), rematerialization (reworking food through technology or decomposition) and revalorization (assigning new or renewed worth or importance to food products through the food chain) are well documented in relation to food transport logistics, food safety and food markets, respectively, they have rarely been studied through an explicit mobilities lens. Equally, the effects of these extrinsic and intrinsic movements of food are far from neutral; they are mediated by entanglements of power and social exclusion that shape how and when people gain access to particular types of food. These entanglements, Cook and Butz (2016, p. 400) suggest, articulate "the intersection of mobility and justice" or mobility justice.

While mobility justice research is dominated by justice issues related to human mobilities (e.g., Sheller, 2015; Vukov, 2015), there is scope to consider its implications in relation to other flows and movements (Hannam, Sheller & Urry, 2006; Urry, 2007), including that of food. Despite productivity gains and more efficient resource use, more than a billion people still experience hunger or nutrient deficiencies, and more than a third of food grown for human consumption is lost or wasted before it is eaten (FAO, 2017). In this context food mobilities rub up against matters of justice with respect to accessing food, which is shaped by the racist, classist and sexist characteristics of the food system that unfairly privilege the wealthy and powerful at the expense of the poor, exacerbating poverty, hunger and food insecurity (Alkon & Agyeman, 2011). Indeed, Dieterle (2015) argues that when food is unavailable and people experience hunger and food insecurity as a result, it is more than a regrettable inefficiency of the industrial food system. It is, in keeping with multiple theorists from Locke to Rawls, an injustice.

I interrogate one contemporary collision of extended food mobilities and social justice concerns that emerges when surplus food that would have previously gone to waste is redistributed to people for consumption. Such redistribution can take diverse forms, including the illicit liberation of edible food from bins and the informal person-to-person gifting of unwanted food. Here I consider the practice of redistributing surplus food from businesses to charities. This focus is timely, as such business-charity surplus food redistribution is a growing activity, particularly within the Global North (EC, 2017; FAO, 2017), stimulated by a suite of social, economic, political, environmental and technological factors. The core question for this chapter is how actors involved in surplus food redistribution – including the donors, recipients and intermediary redistribution initiatives – articulate their goals, practices and impacts at the food mobility-justice nexus. This question is approached first by outlining the current practice of, and governing context for, surplus food redistribution, and then by drawing on empirical evidence from a mobile ethnography of an Information and Communications Technology (ICT)-mediated business-to-charity surplus food redistribution assemblage operating transnationally in both the United Kingdom and Ireland. My analysis attends to the institutional context for surplus food redistribution, incorporating attention to dynamic relations, processes and differential capacities to act that are characteristic of mobility justice analyses (Cook & Butz, 2016). As a result, the chapter makes an empirically grounded contribution to better understanding the logics of surplus food redistribution for human consumption and develops contingent insights into food redistribution's benefits and limitations for simultaneously reducing food waste and food injustice.

The practice and governance of surplus food mobilities

Redistributing surplus food is neither a new nor a homogenous activity. Literature documents evidence of surplus food redistribution through the ages from ancient to contemporary eras (Dikovic, 2016; Hussey, 1997). And my study that maps and analyzes more than 400 surplus food redistribution initiatives across 100 cities around the globe (Davies et al., 2017a, 2017b) demonstrates the heterogeneity of contemporary redistribution processes. Focusing on initiatives that use ICTs (e.g., social media, websites, apps, platforms) to mediate their activities, we found that nearly two-thirds were driven by environmental concerns about food waste and the emissions from landfill it generates. Approximately half of the initiatives focused on providing food for hungry people, while around a third were explicitly committed to achieving both environmental and social outcomes simultaneously, often articulated simply as "connecting businesses that have too much food with charities that have too little" (Food surplus redistribution initiative (RI), Ireland). The flow of surplus food from businesses to charities makes up more than a third of all redistributive activities with retailers, and in particular supermarkets, providing an accessible point to regularly intercept sufficient volumes of surplus food to establish a reliable redistributive stream of food to charities (see also Ciaghi & Villafiorita, 2016). Our study also showed that the majority of the redistribution initiatives that connect donors and recipients are nonprofit organizations. However, within the jurisdiction of the European Union at least, even nonprofit redistribution initiatives are considered "food business operators, placing food on the market" (EC, 2017, p. 6).

Legislation that mediates the flow of surplus food provides an important formal framing of the context within which much redistribution operates and the legal boundaries of practice. These legislative rules have become even more important over the last decade in relation to growing concerns about the level of food waste, what the Food and Agriculture Organization (FAO) calls "a widely recognized global shame" (FAO, 2013, p. 11). Guidelines from the European Commission (EC) state categorically that "[w]hen food surpluses occur, the *best* destination, which ensures the highest value use of edible food resources, is to redistribute these for human consumption" and that "food donation not only supports the fight against food poverty, but can be an effective lever in reducing the amount of surplus food put to industrial uses or sent for waste treatment and ultimately to landfill" (EC, 2017, p. 2, emphasis added). While justice is not explicitly mentioned in these documents, the practice of moving surplus on to feed people in need is presented as feasible and desirable. However, in interpreting both of these statements we are left to imagine the boundaries of the 'value' articulated and the ways in which food donation might support the fight against food poverty, generally defined as the inability to access a nutritionally adequate diet (Dowler & O'Connor, 2012).

Despite its claimed benefits, surplus food redistribution currently mobilizes only a small fraction of available food surplus, with legal and operational challenges facing potential donors and recipients. One key barrier is the infrastructures for food surplus disposal, which are far more established than redistribution processes. Those organizations with surplus food often perceive it to be simpler and cheaper to send food waste to landfill along with the other waste materials they generate. In many places the infrastructures of waste disposal – bins, trucks, dump sites, incinerators – provide a relatively frictionless channel for surplus food to follow, although increasing landfill charges may change this in the future. Equally, disposing of food surplus rather than redistributing it avoids liability risks associated with consuming mishandled surplus food that has been donated. In response, many governments are developing regulations aimed at protecting donors from criminal and civil liability on the one hand to incentivizing donations on the other (FAO, 2011).

While the governing landscape for redistribution is becoming better defined, the justice implications of such enhanced practices have not been the focus of legislation to date. International, national and subnational legislation governing surplus food donation assumes that redirecting surplus food leads to positive outcomes for both food waste reduction and food poverty alleviation if conducted according to food safety guidelines. While a FAO report on food waste notes that "it is not advocated that food donations are the solution to food wastage or poverty", it also claims that "food redistribution can help alleviate the impacts of food poverty. It is the best option in terms of dealing with unavoidable food surplus from environmental, ethic and social perspectives" (FAO, 2013, p. 61). No further details are provided in relation to the substantiation of this judgment, save that "the poorest benefit from nutritious food, and the planet benefits from putting food already produced to its proper use" (FAO, 2013, p. 61). Such sweeping statements are hard to substantiate given the fragmented and poor quality of data on food surplus, which means its nutritional value is poorly understood (Stenmark, Jensen, Quested & Moates, 2016).

Under European law, redistribution organizations are seen as engaging in 'retail', and charities that receive food are considered to be conducting 'mass catering' activities and, therefore, have the same obligations as commercial operators (EC, 2017). Both redistributors and charities are required to record the suppliers of products they receive (one step back) and recipients of the products they redistribute (one step forward), except with respect to the final consumer. This monitoring of redistributed food's journey – its 'traceability' in legislative terminology – occurs under the auspices of technical systems of food risk and safety governance such that the beneficiaries of surplus food redistribution are afforded the same procedural protection as consumers in mainstream marketplaces.

While legislation states that food surplus can help alleviate the impacts of food poverty, the concept of food poverty is rarely well defined in national legislation. In some nations, ratification of the International Covenant on Economic, Social and Cultural Rights (ICESCR) means states have undertaken to tackle poverty, including food poverty, through a human rights framework. This Covenant recognizes that "the roots of the problem of hunger and malnutrition are not lack of food, but lack of access to available food, inter alia because of poverty" (UNESC, 1999, p. 3). Redistributing surplus food, then, provides one potential means of generating improved access, providing the food is of sufficient quality and quantity to meet needs and is provided in "ways that are sustainable" (UNESC, 1999, p. 3). While the UNESC acknowledged in 1999 that realizing the right to food could only be achieved progressively, nearly 20 years later, translating this obligation into practice has remained largely elusive. Indeed, in 2015, the special rapporteur on the right to food stated that "many countries have failed to develop a judicial culture of recognition in practice or the necessary legal frameworks required to ensure that the rights enshrined in the ICESCR are justiciable" (Elver, 2014, p. 2). For example, the EU food donation guidelines make clear that food is to be traceable and edible, but they do not clarify the relative roles and responsibilities of the various actors involved in ensuring that happens. Who provides and pays for the new logistics infrastructures required for the expanded volumes of surplus food redistribution, and who will evaluate the quality of surplus food and its appropriateness for consumption? Without clear answers to these questions, food justice activists and scholars are concerned that lubricating the flow of surplus food through legislation is focused primarily on the needs of the donors, which creates greater risks and dependency for those who are hungry without resolving the underlying causes of either food poverty or food waste (see Caraher & Furey, 2017). The following section provides empirical substance to examine these concerns through a mobile ethnography of surplus food redistribution in Ireland.

Mobile ethnography: following surplus food in Ireland

A mobile ethnography of the assemblage of actors engaging with one surplus food redistribution initiative (RI) in Ireland was conducted between 2015 and 2017. This included formal interviews and informal conversations with key actors – donors, intermediaries, recipients and regulators. These interactions were supplemented with observations of, and participation in, surplus food redistribution in order to examine what happens during the transport of food and at transfer points.

Established in 2012, the RI began as a pilot study between one store and one charity in Dublin and has developed into an international operation connecting more than 4,000 retailers and food industry partners with more than 7,000 charity and community partners across Ireland and the United Kingdom in 2018. Utilizing an app and other forms of ICT (including a website, Twitter feed and text messaging), the RI has redistributed more than 13,500 tons of food, equivalent to more than 28 million meals, for charitable causes, saving them considerable financial expenditure, as well as 43,000 tons of CO_2. How this level of redistribution relates to the total amount of food surplus generated is unclear, for while the Environmental Protection Agency (EPA) estimates food waste in Ireland to be around one million tons per annum, there is no mandatory reporting on what proportion of that is edible. However, voluntary reporting on food waste data by large retailers in Ireland started in 2018, with one multinational already reporting that it donates 17% of its edible surplus annually, with the remainder sent for anaerobic digestion.

In Ireland, the Waste Management (Food Waste) Regulations 2009 specify that most businesses selling or serving food must segregate their food waste at source and that it must not be sent to landfill. There is no reference to redistributing surplus food to people in this legislation. Instead, it is the Food Safety Authority of Ireland (FSAI) that has incorporated the 2017 European Guidelines on food donation in their published short guidance notes explaining food safety standards for charities, donors and redistributors. The Chief Executive of the FSAI has said that the aim of these guidance documents is to help participants comply with food law:

> The FSAI must commend the outstanding work being done by charity organizations and commercial food businesses. ... These partnerships have the best intentions to help people, but with this, there also comes a

great degree of responsibility. Food safety is paramount and all food businesses, including food banks/food redistribution centres, are required to put in place and maintain a food safety management system so they protect the people who ultimately will benefit from the food donation.

(FSAI, 2017)

Although Ireland signed on to the UN International Convention on Economic, Social and Cultural Rights (discussed above), one in eight people remain food insecure (DSP, 2017), and state interventions around food security have long been criticized as limited, fragmented and uncoordinated across a range of government departments (see Dowler & O'Connor, 2012). For example, the updated National Action Plan for Social Inclusion developed by the Department of Social Protection has no food remit, while government visions for agri-food futures are dominated by expansionary plans for commercial food exports (DAFM, 2015) rather than food security. In the following sections I examine the ways in which the RI navigated this policy landscape and addressed the food waste-justice nexus.

Facilitating movement

The RI examined here uses a range of technology innovations, including a dedicated mobile app and an integrated point of sales system (with 'donate' options on barcode scanners used in stores to monitor food flows through the store), to connect retailers with charities and facilitate the donation of surplus food. It is this technological architecture that is often emphasized in media coverage as the reason behind the RI's success. Retailers use an app or web-based platform to post notifications of surplus produce, and charities then receive an automatically generated text message indicating the available food surplus and a time frame in which they must identify what food they require from that surplus.

To comply with legal frameworks around food safety, the RI trains staff, volunteers and charity recipients in food safety, and it augments this training through the use of ICT to ensure full traceability of all donations one step back and one step forward, as required by EU Food Law. This traceability also ensures that organizations which offer poor-quality food, or those that claim food but do not collect it, can be quickly and easily identified, and that relationships can be managed to minimize such instances in the future. ICT then generates reliable data and feedback on impacts and financial gains, which can be an attractive feature to food businesses. Certainly, the RI hopes that the enhanced data on surplus food flows captured through their systems will motivate donors to reflect on their broader purchasing and distribution systems and to reduce persistent and avoidable surplus. This is the RI's main contribution towards food waste prevention, the most desirable option with regards to the management of food in terms of the Irish and EU waste policy frameworks. While there is little evidence that this data shifts retailers' practices, both the technology and relationships formed by the RI are at least opening a space for dialogue in this regard.

Ensuring quality control mechanisms for redistributed food also has been an important motivation for the RI in order to counter charities' negative experiences of poor-quality surplus food donations. One charity recounted their experiences prior to connecting with the RI:

> What [the retailer] was donating, we would have saved them a lot of money. But we ourselves would take on that cost, because we would have to waste all of it. Like say 80% of what they gave us. A large proportion of what they gave us was baked goods, pastries and things, and the rest were vegetables that looked, seriously looked like someone might have stood on them. And why would they donate that? Oh the homeless will eat it. What nonsense.
>
> (Charity 1, Ireland)

Certainly, while the prospect of not having to purchase good-quality food is attractive to cash-strapped community groups and charities, accepting surplus food donations also draws on their limited resources. A common difficulty for charities is the inability to respond to and manage unpredictable offers of surplus food (Caraher & Furey, 2017). In response, the RI moved away from its initial plan in which food was advertised on the app and claimed on a first come first served basis to a more process-driven system that informed charities about scheduling and allocated days for collections and deliveries: "You might think 'oh it's just food coming in the door', but it takes a lot of organization. ... Any little bit of help is great, as long as it's structured" (Charity 2, Ireland).

The RI also has worked with charities to navigate uneven access to and differentiated capabilities with ICT, facilitating a range of basic interactive mechanisms using non-smart phones. In other cases, charities did not have the ability to collect produce from donating stores, so the RI developed a food rescue volunteer operation that manages more than 200 volunteers who collect and deliver food to charities unable to collect it themselves. In 2017, this volunteer operation was supported by Opel, which partnered with the RI by providing seven refrigerated vans to ensure food was efficiently transported and safely preserved in transit. So, while its novel ICT component has generated the most media attention, the RI facilitates a suite of sociotechnical interactions that go far beyond the app to maintain relations between donors and recipients: "If you don't understand the charity and the food business ... you can't just give them an app ... and think oh sure just put it up, somebody will accept it. If you don't have the supporting processes, it does fall down" (RI, Ireland).

The RI's position as an intermediary has not only made donating simpler for retailers; it is also disrupting the traditional power balance between donors and beneficiaries. Acting as an external quality control agent, the RI acts as a filter for poor-quality produce. Moreover, it addresses charities' concerns that if they reject donations, then their future ability to gain access to surplus food might be affected (Tarasuk & Eakin, 2005). Charities are only expected to take what food they have agreed to collect, and they can report substandard foods to the RI, which reduces the potential for food wastage at the charity level and mitigates against charities being used as dumping grounds by powerful multinational supermarkets seeking to make social gains by avoiding landfill costs.

Motivations for participation

Donors and intermediaries overwhelmingly employ instrumental narratives to explain why they participate in the redistribution of surplus food. Indeed, none of them made explicit reference to justice as a motivating force. Even the charitable organizations focused on the ways in which the redistributed food supports their services (which may or may not have a food poverty alleviation agenda) rather than how this resolves any bigger-picture concerns with food justice. Many, however, referred to the importance of the social and relational elements of the redistribution process:

> We were an early partner with [RI] and [retailer], and the difference it has made to our residents is terrific! The food brings the women who live with us together, and we've seen strong friendships develop as they swap and share food and recipes ... not only does this service assist in terms of cost, but it's brought our residents, whether new or long term, together, creating lifelong bonds.
>
> (Charity 4, Ireland)

When asked directly about the social impact of their operations, the RI founders expressed concerns about their ability to accurately establish what difference they are making to the ultimate consumers of the food. As an intermediary between business and charities, the RI has less access to those who consume the food they move around. Ad hoc individual case studies are the main mechanism through which the RI captures some sense of the difference the food makes to consumers. However, they focus mainly on documenting and reporting outputs rather than outcomes, providing the number of charities that participate and the weight of food mobilized rather than relations or capacities. The RI regularly surveys charities to collect data on the operational elements of the redistribution process. This reveals that charities already purchasing food for their clients have reduced their food bills by between 50% and 100%, while other charities have benefitted by adding a food service to their operations for the first time, something that would have been impossible without participating in the redistribution ecosystem. The RI is cognizant of the dangers of making unsubstantiated claims about solving hunger with redistributed surplus food. They acknowledge the social responsibility they create by bringing those in need of food into their food waste reduction initiative. As one founder noted,

> You are creating a dependency from people who are accepting the food, and if you decide after a couple of years that you are either going to sell your technology or get out and move on, then you have built up a dependency and you're kind of gone. And that ... is a lot of responsibility.
>
> (RI, Ireland)

Only on one occasion during the mobile ethnography did a charity talk explicitly about the moral crisis of food waste and hunger, in this case framed through its religious beliefs:

> How could you waste food? And according to the word of the Lord, it's a sin when your brother or sister or community person is hungry and you have this food waste. And then you dump it in the bin when this person is going hungry, cannot feed. That is a sin. So [RI] is a blessing. It's a blessing in Ireland.
>
> (Charity 5, Ireland)

More charities emphasized the importance of ensuring participants are treated in a dignified manner with regard to the quality of food donations:

> Our clients, yes they are in homeless accommodation, and yes they are facing different challenges in life, and they welcome charitable donations. ... But they are still humans and citizens and whatever else. What we were getting in just wasn't good enough. So when RI came along, it was hugely beneficial. ... The food we get is second to none; it's perfect. ... [T]he only issue is it's going out of date.
>
> (Charity 1, Ireland)

A key challenge for the RI in assessing its effect on food consumers is that few charities involved in the redistribution network have the capacity to report on the impacts of the surplus food they receive, outside the cost savings they make as a result of receiving it. The RI, as a result, is increasingly emphasizing its core mission as a logistics intermediary that keeps good food edible and out of the bin, rather than its food poverty alleviation credentials. However, this framing does not mean that its work in moving food around has no justice implications, intended or otherwise.

Surplus food redistribution: attending to the food waste-justice interface

Despite the coalition of forces supporting greater donation of surplus food from retailers to charities, there are also critical voices arguing that such practices may actually work against reducing structural incidences of food surplus and injustice (e.g., Caraher & Furey, 2017). As evidence, they point to the persistence of 'emergency' non-state food banks that have become an enduring feature across many countries, demonstrating that surplus food distribution is making few inroads into reducing food poverty. Critical accounts provide a clear picture of structural failures to eradicate hunger and highlight the dangers of relying on corporate donations without simultaneously holding such corporations accountable for how their business practices – low wages, weak benefits, precarious employment – contribute to food insecurity (Fisher, 2017). Critics rightly argue that both states and corporations have a responsibility to address these underlying causes of food insecurity; the same can be said with respect to addressing the underlying causes of food waste. Participating in surplus food redistribution does not release either states or corporations from these obligations, but does it obscure persistent socially and environmentally unjust practices?

The mobile ethnography found that the RI positions itself as a pragmatic logistics social enterprise working to keep available for human consumption the byproducts of a flawed commercial food system. As one founder said, "the whole thing was around matching food – reducing food waste on one side, and saving costs for charities on the other" (RI, Ireland). They think they have limited capacity to address the underlying causes of food poverty, which requires confronting the operation of the global agri-food industry and structural inequalities related to labor and income. Instead, the RI focuses on acting where it thinks it can achieve tangible, measurable impacts in the short term with respect to reducing food waste, utilizing accessible and effective technology as the means to take immediate action. Narratives around efficiency, safety and value optimization hold center stage. However, whether actively seeking food justice or not, the actions of surplus food redistribution do create a new constellation of benefits and burdens for participating actors that will need to be mapped and tracked over time in order to establish whether they are enhancing distributional justice regarding the right to food. For example, already powerful retailers benefit in multiple ways from the work of the RI; they gain information on the surplus they generate and the surplus redistributed by the RI, make financial savings from landfill tax avoidance and gain public capital by marketing their activities to reduce waste and feed people. Retailers additionally benefit from the increasing protection provided through an expanding governance architecture around food donations that sets out clear requirements for safe redistribution.

Of course, charity recipients also profit from surplus food and their engagement with the RI. If they were already providing food for their clients, then they benefit by reducing food costs. And groups that previously did not offer food have been able to extend their services as a result of connecting with the RI. However, new burdens accompany these benefits; charities become legally responsible for the food they accept – its collection and safe storage, its preparation and its ultimate disposal. Charities unable to shoulder these burdens – those with limited mobility capacities (Cook & Butz, 2016) – are unable to participate. While the RI has responded to these limited capacities in a number of ways, including setting up a volunteer collection system to assist recipients, it too has a limited ability to manage a growing cohort of unpredictable volunteers alongside their existing activities.

Conclusion

In summary, while surplus food redistribution provides greater circulation of edible, safe food to people who are hungry, thereby opening up new channels through which to access food and perhaps contributing to distributional justice on one level, such interventions alone are insufficient to move towards food justice in its broader sense. It is, as Fisher (2017) would say, a 'band-aid' response. Equally, while it keeps more food out of the bin, at least temporarily, more data is required to explore whether the work of the RI and similar initiatives is contributing to a downward trend in the levels of food surplus generated by retailers. To evoke Iris Marion Young (1990), attending to distributional food justice – such as access to food – is necessary but insufficient on its own to develop food mobility justice. Ongoing pressure on retailers and legislators to respond to overproduction and surplus will be required if commitments to food waste reduction by surplus food redistribution initiatives are to be achieved. Greater dialogue with recipients around the impacts of participating in surplus food redistribution is also required. Certainly, unjust food mobilities, with their entanglements of power and social exclusion, will not be resolved with technical fixes alone. Much deeper engagement with the sociopolitical realities of the global food system and its links to food waste and poverty are required, and not only by grassroots organizations at the coalface of waste and hunger. Echoing Mullen and Marsden (2016), mobility justice in relation to surplus food redistribution cannot be reduced to technical questions around the scale of access or speed of distribution. Greater collective thinking about how food should be grown, moved, accessed and used is required, as well as greater dialogue among scholars working on food, waste, mobilities, social practices and justice.

Mobility justice provides a useful way to talk about the different relations that structure the movement of surplus food, highlighting power differentials that shape that practice and provide the context within which current food waste and food injustice collide. Schemes that rely on voluntary donations collected and redistributed by an assemblage of non-state organizations do not in isolation offer a long-term, reliable source of food; nor do they necessarily hold governments to account with respect to their obligations around ensuring a right to food. To avoid becoming part of the problem of food injustice, redistribution initiatives must ensure they continue to monitor and report on their impacts, and they must also call for upstream action from both states and corporations to address unsustainable food systems.

Acknowledgments

This chapter draws on empirical research from the SHARECITY project that was funded by the European Research Council under the European Union's Horizon 2020 research and innovation programme (grant agreement No 646883).

References

Alkon, A. H. & Agyeman, J. (2011). *Cultivating food justice: Race, class and sustainability.* Cambridge, MA: MIT Press.

Caraher, M. & Furey, S. (2017). *Is it appropriate to use surplus food to feed people in hunger? Short-term band-aid to more deep-rooted problems of poverty.* Food Research Collaboration Policy Brief. London: London City University.

Ciaghi, A. & Villafiorita, A. (2016). 'Beyond food sharing: Supporting food waste reduction with ICTs'. *2016 IEEE International Smart Cities Conference (ISC2) proceedings* (pp. 1–6). doi:10.1109/ISC2.2016.7580874.

Cook, N. & Butz, D. (2016). 'Mobility justice in the context of disaster'. *Mobilities*, 11 (3), 400–419.

DAFM. (2015). *Food Wise 2025: Local roots, global reach.* Dublin: Department of Agriculture, Food and the Marine. Retrieved from www.agriculture.gov.ie/foodwise2025.

Davies, A. R. (2012). 'Geography and the matter of waste mobilities'. *Transactions of the Institute of British Geographers*, 37(2), 191–196.

Davies, A. R., Edwards, F., Marovelli, B., Morrow, O., Rut, M. & Weymes, M. (2017a). 'Creative construction: Crafting, negotiating and performing urban food sharing landscapes'. *Area*, 49(4), 510–518.

Davies, A. R., Edwards, F., Marovelli, B., Morrow, O., Rut, M. & Weymes, M. (2017b). 'Making visible: Interrogating the performance of food sharing across 100 urban areas'. *Geoforum*, 86, 136–149.

Dieterle, J. M. (2015). *Just food: Philosophy, justice and food.* London: Rowman & Littlefield International.

Dikovic, J. (2016). 'Gleaning: Old name, new practice'. *The Journal of Legal Pluralism & Unofficial Law*, 48(2), 302–321.

Dowler, E. & O'Connor, D. (2012). 'Rights-based approaches to addressing food poverty and food insecurity in Ireland and UK'. *Social Science & Medicine*, 74(1), 44–51.

DSP. (2017). *Social inclusion monitor 2015.* Dublin: Department of Social Protection. Retrieved from www.socialinclusion.ie/documents/2017-05-18_SocialInclusionMonitor2015_rpt_FinalWcov_000.pdf.

EC. (2017). *EU guidelines on food donation* (2017/C 361/01). Brussels: European Commission. Retrieved from https://eur-lex.europa.eu/legal-content/EN/TXT/?uri=CELEX:52017XC1025(01).

Elver, H. (2014). *Access to justice and the right to food: The way forward* A/HRC/28/65. Geneva: Human Rights Council, General Assembly.

FAO. (2017). *The future of food and agriculture.* Rome: FAO. Retrieved from www.fao.org/3/a-i6583e.pdf.

FAO. (2013). *Toolkit for reducing food wastage footprint.* Rome: FAO. Retrieved from www.fao.org/docrep/018/i3342e/i3342e.pdf.

FAO. (2011). *Global food losses and food waste: Extent, causes and prevention.* Rome: FAO. Retrieved from www.fao.org/docrep/014/mb060e/mb060e00.pdf.

Fisher, A. (2017). *Big hunger: The unholy alliance between corporate America and anti-hunger groups.* Cambridge, MA: MIT Press.

FSAI. (2017). 'FSAI provides food donation guidance to charities and businesses'. *Food Safety Authority of Ireland.* Retreived from https://www.fsai.ie/news_centre/press_releases/food_donation_29092017.html.

Gibson, S. (2007). 'Food mobilities: Traveling, dwelling and eating cultures'. *Space & Culture*, 10(1), 4–21.

Hannam, K., Sheller, M. & Urry, J. (2006). 'Mobilities, immobilities and moorings'. *Mobilities*, 1(1), 1–22.

Hussey, S. (1997). 'The last survivor of an ancient race: The changing face of Essex gleaning'. *The Agricultural History Review*, 45(1), 61–72.

Mullen, C. & Marsden, G. (2016). 'Mobility justice in low carbon energy transitions'. *Energy Research & Social Science*, 18, 109–117.

Sheller, M. (2015). 'Racialized mobility transitions in Philadelphia: Urban sustainability and the problem of transport inequality'. *City & Society*, 27(1), 70–91.

Stenmark, A., Jensen, C., Quested, T. & Moates, G. (2016). *Estimates of European food waste.* Stockholm: IVL Swedish Environ Res Institute.

Tarasuk, V. & Eakin, J. M. (2005). 'Food assistance through "surplus" food: Insights from an ethnographic study of food bank work'. *Agriculture & Human Values*, 22(2), 177–186.

UNESC. (1999). 'International covenant on economic, social and cultural rights'. General Comment No. 12. Committee on Economic, Social and Cultural Rights. Geneva: Office of the United Nations High Commissioner for Human Rights.

Urry, J. (2007). *Mobilities.* Cambridge: Polity.

Vukov, T. (2015). 'Strange moves: Speculations and propositions on mobility justice'. In L. Montegary & M. A. White (Eds.), *Mobile desires: The politics and erotics of mobility justice*, pp. 108–121. Basingstoke, UK: Palgrave McMillan.

Young, I. M. (1990). *Justice and the politics of difference.* Princeton, NJ: Princeton University Press.

Index

Printed in the United States
by Baker & Taylor Publisher Services

Printed in the United States
by Baker & Taylor Publisher Services